ROCK STAR

ROCK STAR

*The story of Reg Sprigg –
an outback legend*

KRISTIN WEIDENBACH
Best-selling author of *Mailman of the Birdsville Track*

East Street Publications

Published by East Street Publications,
PO Box 307
Hindmarsh
Adelaide SA 5007
South Australia

www.eaststreet.com.au

Copyright © Kristin Weidenbach 2008

All rights reserved. Without limiting the rights, under copyright above, no part of this publication shall be reproduced, stored in or introduced into a retrieval system, or transmitted in any form or by any means (electronic, mechanical, photocopying, recording or otherwise), without the prior permission of both the copyright owner and the publisher of this book.

First published 2008

Cover images:
Front cover
Main image—Geosurveys Hill, Simpson Desert, 1964.
Fossil image—*Dickinsonia costata*
1. Two-year-old Douglas helps Reg fix a broken axle at Mungerannie on the Birdsville Track, 1956.
2. Geltwood-1 drill rig, Geltwood Beach, SA.
3. Reg on the Simpson Desert crossing, 1962.

Back cover
Arkaroola

National Library of Australia Cataloguing-in-Publication entry

Author:	Weidenbach, Kristin.
Title:	Rock star : the story of Reg Sprigg - an outback legend / Kristin Weidenbach.
Edition:	1st ed.
ISBN:	9781921037290 (pbk.)
Notes:	Includes index.
	Bibliography.
Subjects:	Sprigg, Reg C. (Reginald Claude)
	Arkaroola–Mt Painter Sanctuary.
	Geologists–Australia–Biography.
Dewey Number:	551.092

Cover design by Ellie Exarchos, Scooter Design
Map created by DEMAP
Design and typeset by David Bradbury
Typeset in 12/16 Garamond
Printed in China at Everbest Printing Co.

*To Reg's grand-daughter—Lila Rose Sprigg Weidenbach
My daughter. My greatest joy.*

Three-year-old Lila at Ediacara South, the site of Reg Sprigg's 1946 Ediacaran fossil discovery, December 2007. *(Kristin Weidenbach)*

Note
Australian pounds (£), shillings (s) & pence (d); and dollars ($) & cents (c) have been used throughout the text to reflect the period in which they were in use. Australia transferred to decimal currency on 14 February 1966.

What a delight it is to escape this cruel old world of ours and converse with nature. I could wander for hours without a care in the world, but always finding something new. What a privilege to be geologically inclined—and what an escape!

Reginald Claude Sprigg, 22 March 1994

Rugoconites enigmaticus

Contents

List of maps and illustrations — IX

Acknowledgements — XIII

Preface — XV

- CHAPTER 1 Early Days — 1
- CHAPTER 2 Wartime Uranium Search — 24
- CHAPTER 3 The 40-mile-an-hour Geologist — 47
- CHAPTER 4 Radium Hill — 63
- CHAPTER 5 Mapping the State — 85
- CHAPTER 6 Oil — 111
- CHAPTER 7 Across the Simpson Desert — 140

CONTENTS

CHAPTER 8 Waterworlds — 169

CHAPTER 9 Success in the 1960s — 193

CHAPTER 10 An Outback Wildlife Sanctuary — 213

CHAPTER 11 Bad Times — 239

CHAPTER 12 Battered Beach — 262

CHAPTER 13 One Lifetime is Not Enough — 285

CHAPTER 14 The Sprigg Legacy — 308

References — 325

Index — 328

Charniodiscus arboreus

List of maps and illustrations

Reg's granddaughter — v
Reg Sprigg's primary area of operations — XVIII
Field trip with Mawson and Madigan — 12
Mawson packing geological specimens — 20
Road-making in East Painter gorge, Arkaroola — 31
Bentley Greenwood driving at Arkaroola Station — 34
Reg with drill cores — 57
Griselda — 65
Reg, Menzies and Playford in Radium Hill mine — 68
Reg and Griselda, wedding — 80

LIST OF MAPS AND ILLUSTRATIONS

Griselda at Radium Hill — 88
Spinifex mounds at Mount Davies — 102
Santos's Wilkatana oil field — 126
Anticlines — 136
Santos Chairman John Bonython — 138
Geosurveys dining bus — 143
Reg and children, Coonana bore — 144
Playford at Innamincka-1 well site — 149
Reg in underwear, Simpson Desert — 157
Margaret and Douglas with Poeppel Post — 165
Reg and Darby, meal-break in the Simpson — 167
MV *Saori* — 176
Darby enters the diving chamber — 184
Arkaroola house circa 1967 — 218
Arkaroola house, 2007 — 219
Sillers Lookout — 237
Helicopter gravity survey — 242
Reg and Dame Joan Sutherland — 264
Arkaroola village — 281
Reg's 1947 diagram of Murray submarine canyons — 317
2005 satellite image of Murray submarine canyons — 317
Geological timeline — 324
Author photo — 334

List of plates
Griselda and children at Baker street home
Premier Don Dunstan at Arkaroola
Reg receives the Order of Australia
Reg with young family
Field trip with Mawson to the Flinders Ranges
Mundrabilla meteorite

Reg drives over a Simpson Desert sand dune
Lady Paquita Mawson opens Arkaroola's Mawson lodge
Reg at home at Arkaroola
Reg and Sir Mark Oliphant
Reg and Griselda, Simpson campsite
Santos oil consultant, AI Levorsen
Geosurveys Hill

List of Ediacaran fossils

Rugoconites enigmaticus — VIII
Charniodiscus arboreus — X, 24
Tribrachidium heraldicum — XIII, 239
Spriggina floundersi — XV, 213
Beltanella gilesi — 1
Ediacaria flindersi — 47
Cyclomedusa — 63
Dickinsonia costata — 85
Dickinsonia lissa — 111
Chondroplon bilobatum — 140
Mawsonites spriggi — 169
Parvancorina minchami — 193
Eoporpita — 262
Funisia dorothea — 285
Inaria karli — 308
Marywadea ovata — 325
Phyllozoon hanseni — 328

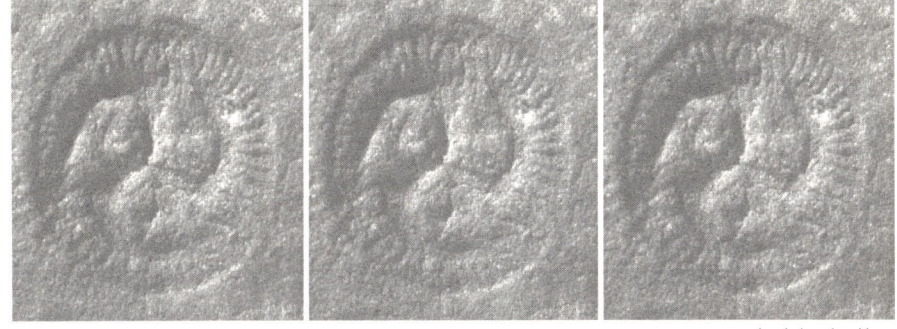

Tribrachidium heraldicum

ACKNOWLEDGEMENTS

FIRSTLY I'D LIKE to thank Reg, himself, for having the interest and foresight to build a personal archive of his life. He was a man who appreciated history and valued the preservation of thoughts, opinions and ideas. He kept copies of almost every letter, speech or essay that he wrote. Without his extensive collection of personal papers, this book would not have been written.

For allowing me access to Reg's world I sincerely thank his children, Marg and Doug Sprigg, who allowed me to immerse myself in Reg's memory without restriction. All of the photos in the book are from the Sprigg family collection unless otherwise stated.

I would also like to express my gratitude to Reg's friends and colleagues who granted me interviews. To Allan Vial, Betty Wilson, Bob Laws, Chris von der Borch, Clyde Thomson, Jean Shaw, Keith Johns, Lee Parkin, Ngaire Teesdale-Smith, Norton Jackson, Reg Nelson, Scoresby Shepherd, Trish

Wright, Rowl Twidale, Dorothy Francis and Wolfgang Preiss, your personal memories and anecdotes have greatly enriched my book.

I would also like to thank Jim Gehling, Rick Wilkinson and Ian Plimer for general assistance and encouragement, along with Lorraine Edmunds, Mark Pharaoh and Doug for offering comments on the manuscript.

Thanks to Jim and the South Australian Museum for supplying Ediacaran fossil images. Thanks also to Helen Bruce of the University of Adelaide Archives; to Cheryl Hoskin, Special Collections Librarian at Adelaide University's Barr Smith Library; to Kate List and Paul Jamieson from Geoscience Australia; to Vivian Wilson for permission to publish from the Oliphant collection, and to the family of Richard Grenfell Thomas for permission to publish from RGT's letters and photographs.

I am grateful to Bernie O'Neil and Barry Cooper for their extensive interviews with former South Australian Department of Mines and Energy staff—including Reg—transcripts of which have been lodged in the State Library of South Australia and which I relied upon greatly. Thanks also to Nancy Flannery for access to the interview she conducted with Reg in September 1994.

Thank you to my publishers Michaela Andreyev and Jane Macduff for their excitement and general enthusiasm for all things Sprigg. I believe we have made a book of which we can all feel very proud.

And of course, thank you to my family and close friends who have supported me in so many ways throughout the writing of this book. I am truly blessed.

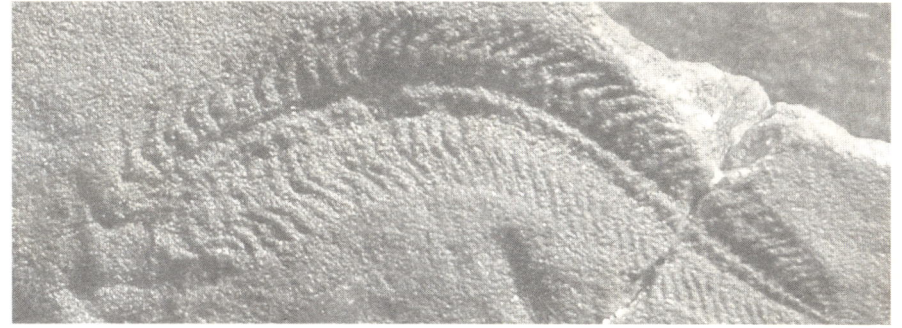

Spriggina floundersi

Preface

On the 23 April 1990, I was a 23-year-old biology student graduating from Flinders University. Sitting in the Adelaide Festival Theatre in black academic gown and ceremonial purple hood I was just one of hundreds of students receiving my degree that day. At the conclusion of all the standard degrees the vice-chancellor presented one special degree—Doctor of Letters *honoris causa*—to a balding older man in crimson gown and flat, black, velvet hat. Like most of the young graduates I listened with one ear, fidgeted in my seat and waited for the long ceremony to be over while this newly conferred doctor made his acceptance speech. My mother, sitting further back in the audience, said it was an absorbing account of seafaring ancestors and adventures in scuba diving and underwater exploration. How was I to know then, that I would one day fall in love with this man's son.

That I would give birth to his grand-daughter. That I would find his life so fascinating that I would want to write a book about it.

Reg Sprigg was Australia's quixotic geologist. More visionary than Sir Douglas Mawson; more entrepreneurial than Cecil Madigan; more successful than Sir Edgeworth David. An oil field is named after him as well as a fossil worm (*Spriggina floundersi*), a radioactive mineral (spriggite), an asteroid (Asteroid Sprigg, discovered 7 May 1991 by RH McNaught at Siding Spring), a submarine canyon (Sprigg canyon), a scholarship for Aboriginal and Torres Strait Islander undergraduate students (Adelaide University's Reg Sprigg Scholarship), a prize for Australia's best oil or gas crusader (Australian Petroleum Production and Exploration Association's Reg Sprigg medal), and a coastal inlet on the north-east tip of Kangaroo Island (Sprigg inlet) among other things.

Charismatic and entertaining, he made science come alive. When Reg talked about the Ice Ages you could imagine the poles expanding, the oceans being sucked into ice sheets and the newly exposed beaches swirling with eddies of dust. When he talked about crustal geology the rocks became fluid things that squeezed and pushed, anticlines mounded up and synclines sank, reefs of gold snaked into the ground and arched back out again. When he talked about fossils you could see sinuous segmented worms wriggling in the shallows and curious jellyfish-like discs settling in the sand; you could feel the tepid stillness of the water and hear the flat silence of a world without birds or crickets, people or trees.

Reg Sprigg was a man of contrasts. He devoted his life to the wonder of the natural world yet moved with ease in the corporate world of business and politics. He focused his quick intelligence on elucidating the mysteries of the earth yet effortlessly absorbed history, art, music and literature as he progressed on his eternal voyage of discovery. He successfully guided his family and followers on the first motorised trip across the trackless Simpson Desert yet couldn't remember where he'd parked his car in the city and rode around and around in a taxi looking for it. He dined with prime ministers, premiers and governors yet went to deliver a lecture at Adelaide University forgetting to wear his socks. He could negotiate complex business deals with

Griselda, Margaret and Douglas in front of their Baker Street home, March 1957.

South Australian Premier Don Dunstan and his first wife, Gretel, visited Arkaroola in August 1970.

Reginald Claude Sprigg receiving his Officer of the Order of Australia medal from South Australian Governor Sir Donald Dunstan, at Government House, Adelaide, 20 September, 1983.

Reg cradles newborn Douglas (four weeks old) while Griselda holds two-and-a-half-year-old Margaret, December 1954.

Members of the Adelaide University final year geology class, 1940, atop Wilyerpa Hill, southern Flinders Ranges, June 1941. Shirley Edwards and Reg, standing at top; Sir Douglas Mawson, standing at right; K Machell, Gordon Haskard and John Shepherd, seated (left to right).

Griselda with the smaller, 5-ton fragment of the Mundrabilla meteorite found by Geosurveys employees Bruce Wilson and Tony Cooney in March 1966. Together with the larger, almost 12-ton fragment, it was the largest meteorite discovery ever made in Australia. Geosurveys Woodville depot, April 1967.

Reg guides the Toyota over a sand dune during the Sprigg family's third crossing of the Simpson Desert, August–September 1967.

Lady Paquita Mawson, widow of Sir Douglas, opening Arkaroola Wildlife Sanctuary's Mawson accommodation lodge, June 1970, flanked by Reg, Griselda and Margaret Sprigg.

Reg on the verandah of his Arkaroola home, March 1993. He named Griselda Hill (in the background) for his wife when she made her first ascent in 1968.

Two dear friends—Reg Sprigg and Sir Mark Oliphant—unveiling the commemorative cairn for Mt Oliphant (in background), Arkaroola, March 1986.

Reg and Griselda rest under a make-shift shelter on T19 claypan near Geosurveys Hill in the centre of the Simpson Desert, September 1964.

Reg consults the map with Californian oilman Neil Willis (left) and American oil consultant to Santos, Al Levorsen (centre), on a bush airstrip in Central Australia, November 1958.

Reg and Douglas at the cairn atop Geosurveys Hill, the only hard crustal hill in the Simpson Desert, September 1967.

PREFACE

international oil companies yet lose his favourite hat in the bottom of his wastepaper basket while the whole office searched madly to find it. He spent his life mining and exploring yet dedicated his resources to conservation and environmental protection. He spent more time away from his family than with them yet his death created a void that time has not filled.

Reg Sprigg is regarded by many as the greatest geologist of his generation and one of the best Australia has produced. By the age of thirty he had discovered the oldest fossils in the world and some of its deepest undersea canyons. He had worked at Australia's first two uranium mines and searched for material with which to construct the world's first atomic bomb. By the time he was 40 he had set up the largest private geological consulting firm in the country, helped found Santos, and discovered the great Cooper Basin oil and gas fields. By the time he was 50 he had built his own boat and diving bell and mastered the primitive art of scuba diving. He had driven the first vehicle across the Simpson Desert and crossed the continent from north, south, east and west. He had also launched Arkaroola Wildlife Sanctuary, one of Australia's first ecotourism resorts. The rest of his life he would devote to the unlikely bedfellows of mining and conservation, and to his insatiable love of geology.

People who knew Reg speak of his fun-filled manner and his zest for life. He was a man filled with big ideas and sweeping vision. His confidence and enthusiasm were infectious, and articulating his dreams inspired others to join his quest. He was always at the forefront of his field and only now is the rest of the world coming to appreciate the extent of his endeavours. This is his story.

ROCK STAR

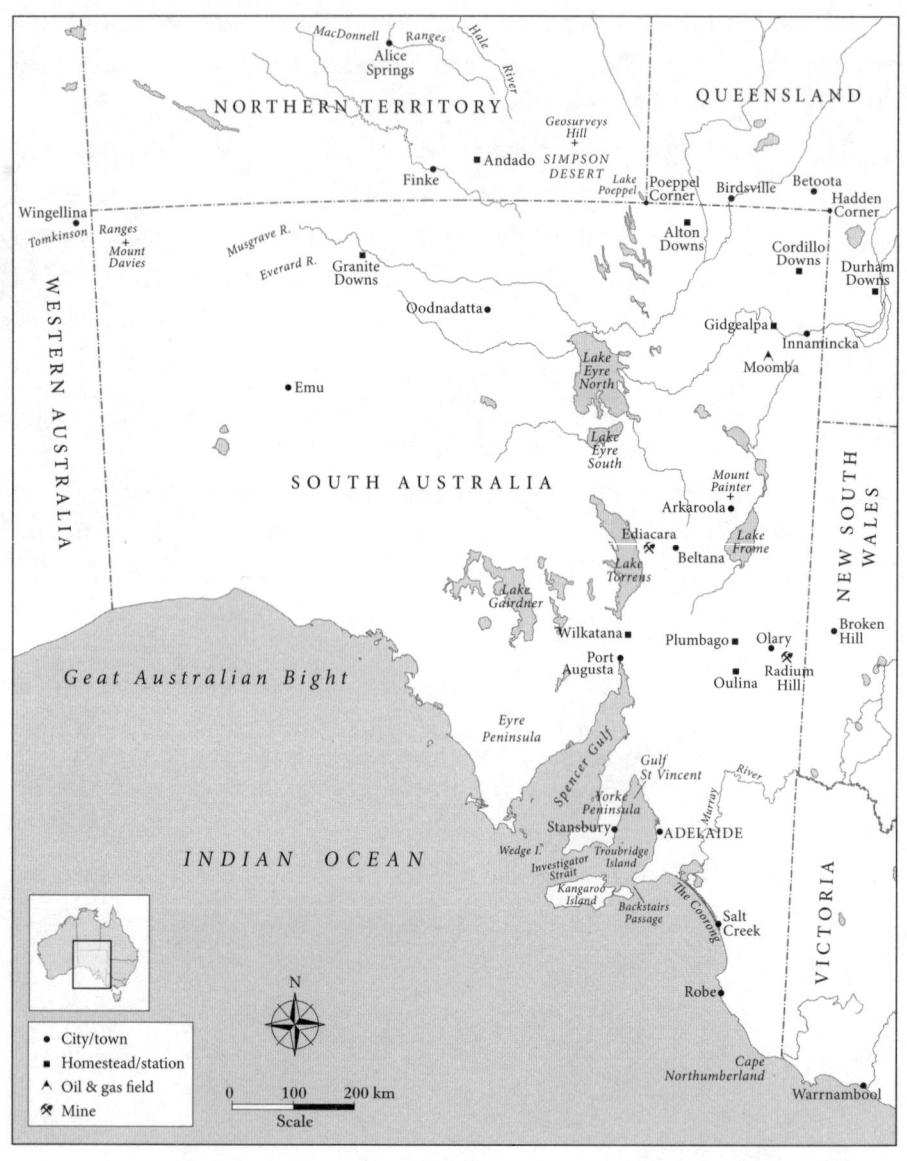

XVIII

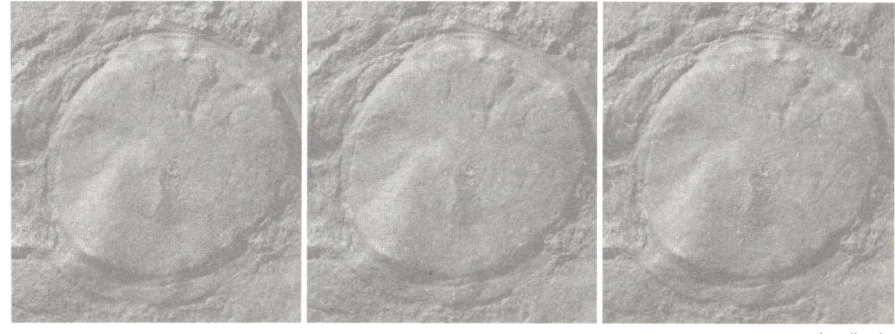

Beltanella gilesi

CHAPTER 1

EARLY DAYS

YOUNG REGINALD CLAUDE Sprigg came into existence on the 1 March 1919. It was less than four months since Armistice Day and the world was still shedding the grim cloak of war. Scores of young men from the small town of Stansbury went to fight in the battles of Europe. Too many of them didn't come home.

According to family lore Reg was a small, premature baby, not given much chance of survival. 'My mother was advised not to have any more children after two and she was suddenly carted off to the Yorketown Hospital and had me on the side of the road,' as he later explained it. His unceremonious beginning didn't stop him from thriving. In fact it was older brother D'arcy who was always the sickly one; who died at 37 from cancer.

Reg was born into a ready-made family of four, joining sister Constance Vera (Connie), born in 1916, and brother, five-year-old D'arcy Kingsley, born in 1914. The Sprigg children lived at Stansbury, on the Achilles tendon of South Australia's Yorke Peninsula, with their mother Pearl and their father Claude. It was where Reg's maternal grandfather had settled in the late 1800s and fathered nine children, a procession in which, Pearl Alice Irene Germein—Reg's mother—landed right in the middle.

In looks, Reg resembled his maternal Germein ancestors. Their physical features as well as their physical presence and family influence, dominated his early childhood. The Spriggs, in contrast were rarely mentioned, and only then by his mother in muttered, disapproving tones. One got the feeling that the Spriggs were rather not good enough for the dignified Germeins.

Reg's father, Claude Augustus Sprigg, was the singular exception, in Pearl's eyes. He had washed up in the coastal township one day like a sand-dollar thrown up by the sea. Captain John Germein's eldest daughter developed an affection for the young Sprigg—four years her junior—and enveloped him into the Captain's brood. Naturally he was much better off in the Germein family bosom, she surmised, with all thought of the Sprigg side of things best forgotten.

There was never any explanation for this rejection. Young Reg grew up believing that his paternal grandparents were long dead and the only thing Sprigg in their family was the surname. He waited 70 years to learn that his grandfather actually died in 1935. He'd had a grandpa for 16 years that he hadn't known he had. By the time Reg found out, however, both his parents were also dead and there was no one left to ask. No explanation for why Gus Sprigg had been excluded from their lives. The only potential clue was found in the yellowed newspaper notice of his death:

'Mr Samuel Augustus Sprigg, an old age pensioner and a long resident of the Naracoorte district, died in the Naracoorte hospital on Friday, where he had been a patient for some time,' *The Naracoorte Herald* announced. 'The late Mr Sprigg followed farming and grazing successfully at one time, and held a property near Naracoorte, but in latter years he met with misfortunes and did not enjoy a full measure of life's goods. He had a good disposition, and was well informed on most subjects, and possessed ability also, belonging to a family who were respected and esteemed in this district where they resided some years ago.'

The 'misfortunes' that Samuel Augustus Sprigg met with that prevented him enjoying 'a full measure of life's goods,' were not elaborated upon, but Reg suspected that it was these mysterious misfortunes that had lead to grandfather Sprigg's presence being reduced to a vague, shadowy outline of which his mother dismissed any mention with primly pursed lips.

Perhaps the misfortunes had begun when Augustus's wife had died suddenly following the birth of their fourth child. In the morning she had been quite fine. By two o'clock in the afternoon she was dead. 'The neighbours around rendered every assistance that lay in their power, but without the aid of proper nursing and medical assistance the case proved hopeless,' reported the newspaper of 29 April 1902. The Frolichs were a well-known and respected family of early settlers in the district and the township of Naracoorte lamented young Charlotte Dorothea's loss. She was only 30 years old.

Twelve-year-old Claude Augustus Sprigg didn't linger at home after his mother's death. He made his way to Stansbury to reside with his brother, and 11 years later married his Pearl.

By the time Reg came along, Claude—Pop to his children—was a storekeeper at the Stansbury General Store. Later, Pop bought the store and worked day in and day out to build up his business. Pearl began to worry about his health. He was working himself into the ground, she complained.

For Reg, though, it was a normal happy childhood in the company of siblings and cousins. His grandfather, Captain John, who lived on a farm just north of town, entertained them all with stories of seafaring days—his own and those of his father, grandfather and uncles. The Germeins were genuine original pioneers, he informed young Reg proudly. His grandfather, the original Captain John Germein, had been brought out to the new colony by the South Australian Company to survey the local waters and take charge of trawling and fishing for the new settlement at Kangaroo Island.

'...we trust you will show, by your conduct and disposition, how much you desire to see prosperity and success attend this very important expedition,' George Fife Angas, Chairman of the South Australian Company had written to Germein on his departure from Plymouth in November 1836. 'As master of a trawler you will avoid all fishing on the Sabbath day, and we trust you will do all you can to assist in teaching in the schools on that day, thereby communicating to others the knowledge you have derived from the same kind of instruction in your native land. Wishing you and your family every good, and trusting to Divine Providence for a safe and prosperous voyage.'

The Captain's three sons became expert seamen at his side. John Germein II as master of the government cutter *Waterwitch*, was dispatched to the head of Spencer Gulf in 1840 to convey provisions and stores to Edward John Eyre for his expedition to the centre of Australia. On the way back he explored the eastern shore of the gulf and discovered 'a good harbour with five fathoms water.' South Australia's Governor Gawler named it Germein's Roadstead in his honour—later to become the township of Port Germein.

John's brother Samuel, as master of *Hero*, similarly supplied Eyre on his journey to western Australia.

The third brother, Benjamin, was renown for his skilful piloting—rowing out to meet incoming windjammers and steamships and piloting them to safe anchorage at the Semaphore heads. Other seafarers acclaimed his feats in handling vessels up to 600 tons 'like yachts,' bringing them right up the Port River, without buoys, beacons or dredged channels. He was particularly known for the special attention he gave to vessels under canvas, and his love of the pilot service began to diminish when the era of sail gave way to steam. 'Any dredge master or hopper skipper could navigate the river in a steamer,' he'd say. 'It wanted a smart fellow to work up a big sailing craft.'

Ben's eccentricity began to increase as his self-perceived usefulness decreased. Often he would take a boat out alone on the waves, but being such a 'good man in a boat' his absence for a few days was not concerning. Then he started leaving suicide notes behind and disappearing for several weeks. On the 3 July 1893, he left home, and when nothing had been seen

EARLY DAYS

of him by the end of August, the police were notified. It was not until near the end of September that the water police found his decomposed body laying face down among the mangroves at high water mark about a mile from the Port River. 'A glass bottle filled with sand was lashed round the deceased's neck, a heavy winch handle was tied round the waist, and a bag of stones was attached to the feet, which were bound together with a bit of lashing,' *The Adelaide Observer* of 23 September 1893 reported. It seemed that this time pilot Germein had really meant it.

John Germein III, Reg's grandfather, grew up in the seafaring tradition. He had stories to tell of helping on his father's ship, *Omeo*, trading between Port Pirie—the neighbouring harbour to Port Germein—and Wallaroo, further down the gulf, where the copper mines were driving a flourishing local economy; and stories too, of joining a crew sailing to China at 21 years old and jumping overboard during a typhoon to rescue a drowning shipmate. But in the end he decided to settle closer to home and assumed control of the steamer *Ceres*, which was the main means of transport from Port Adelaide across Gulf St Vincent to the Yorke Peninsula. He captained *Ceres* for 27 years and made 4000 trips carrying passengers and cargo to the east coast peninsula ports of Stansbury and Port Vincent.

When his seafaring days began to wane he bought a farm just out of Stansbury and turned his attention to the land. His last maritime adventure was jumping from the end of the Stansbury jetty at age 65 to rescue six-year-old Victor Ralph Norsworthy from floating out to sea: A feat for which he received an award from the Royal Humane Society.

Rescuing obviously ran with salt water through the Germein veins. When uncle Benjamin had been the first lighthouse keeper at Cape Northumberland, near the border of South Australia and Victoria, he had lost some of his fingers rescuing the captain and two surviving crewmen from the capsized brig *John Ormerod*. The fingers had been crushed between his boat and the brig as he ranged up alongside in heavy seas. Two years earlier, in August 1859, he had proved his saviour mettle when he had rescued the final survivors from the wreck of the *Admella*. The steamer was on her final trip from Adelaide to Melbourne with 93 tons of cargo, seven horses, 84 passengers and a crew of 29 aboard when she became impaled on Carpenter Rocks. Lighthouse keeper Germein salvaged a small boat that had washed ashore from the *Admella* by patching it up with soap and rope-yarns, selected a scratch crew from the bystanders on the shore, and made five attempts to reach the survivors, still—after seven days—clinging to the remnants of the steamer.

'Three attempts were foiled by the gigantic breakers capsizing their indifferent craft, and they themselves only regained the shore with utmost difficulty by God's providence, to courageously make yet other attempts,' read an account of the rescue, 'until on the fourth and fifth essay Germein succeeded in laying his boat under the stern of the wreck and taking off several persons including the Captain...After safely negotiating the reef breakers, and nearing the shore surf, the terrible waves upended their boat and they were all immersed in the breakers some 100 yards from the shore. Despite the most gallant exertions of Germein he was unable to save one poor unfortunate, who was thus drowned literally in sight of rescue. Being a strong swimmer, Germein was able to help the others through the surf to safety and the succour of the shore helpers.'

In all, 89 lives were lost in one of Australia's worst maritime disasters.

Ben Germein's intrepidness earned him the gold medals of the Royal Humane Society and the Shipwrecked Mariners' Society, in addition to one of the silver medals specially struck by the governments of South Australia and Victoria to commemorate the courage of those involved.

THUS THE GRANDSON of Captain John Germein had a proud maritime heritage, but for young Reg Sprigg, the stories of shiplife and bravery were just fireside tales. His was a country-town life; that is until Pop sold the Stansbury store and the family moved to Adelaide. Reg was only five years old and the only thing he really remembers missing is visits to the town beach to collect interesting treasures. His favourites were pretty, white, heart-shaped 'pebbles' with a lacey star pattern on their backs, which he later found out were fossil sea urchins called *Lovenia*. Fan shells and the smooth grey ovals that the locals called "parrots' beaks," (fossil lamp-shells) also took his fancy. These gems littered the cobbled beach and could be easily scraped out of the limestone cliffs lining the foreshore with a stick and transported home in bulging pockets. The five-year-old discovered a primordial thrill of collecting that stayed with him for life.

When the family moved to Goodwood, in the city, Reg composed his first letter, offering one of his Stansbury cousins a penny a piece for as many of the lovely *Lovenia* as he could find. No response. It was left to Reg to expand his collection himself on family vacation trips back home. These were achieved firstly via steamships or ketches to make the 50 kilometre crossing of the gulf, then latterly via Pop's brand new 1926 Chevrolet Tourer.

The first trip by car was in the winter of 1927, after the Chev had been 'run in' and Mother had had sufficient time to feel at ease with Pop's driving. He felt himself quite dapper coasting along with his new eye goggles, long scarf and neatly trimmed moustache. Mother soon pulled him into line, however, if his new mobile status went to his head. 'You're a speed maniac Father!' she yelled at him one day as they raced along Anzac Highway at almost 48 kilometres per hour. 'You'll have us all killed! I don't know what has come over you since you bought this car. I wish I'd never seen the damn thing.'

It was six o'clock in the morning when the family set off on the Chev's inaugural Stansbury run. They had a long day ahead of them—100 kilometres to the top of the gulf and another 100 kilometres down the other side, along an unsealed track winding across marshlands and sand dunes, and littered all the way with skittery, round limestone pebbles and puncture-forming discarded horseshoe nails. The Chev's running boards were loaded with petrol cans, water and all kinds of provisions. The spare wheel, tyre patches and spare tubes were soon in demand. The first two punctures came just after Two Wells—only 40 kilometres from home.

It was nine o'clock at night—15 hours to drive 200 kilometres—before they finally reached Uncle Syd's farm after 19 punctures and several boggings, from one of which they had to pay a farmer with a team of draught horses ten shillings to extricate them. A scandalous price! Pop declared. Mother meanly suspected the enterprising farmer of deliberately wetting the track to improve his earnings from stranded motorists like themselves.

VISITS TO THE peninsula to improve his fossil collection were fun times but Reg did not particularly pine for the coast. He had taken the move from Stansbury in his stride and settled into city life with ease. For a while the family lived with Great Aunt Martha at Goodwood and Reg attended the local primary school. The only thing he didn't like was going to Sunday School. Some mornings his pre-religious excavations in the back yard fouled up his good clothes so much that his exasperated mother agreed he really *couldn't* go to church looking like that. It wasn't really deliberate sabotage on Reg's part. The trench he was digging in the backyard was simply too mesmerizing. Half a metre below the surface a layer of pale limestone appeared in the soil, which was then interspersed with layers of hard red clay down to three metres or so—the limit of his bucket-raising strength and thus the bottom limit of his hole. He loved the way the material changed as he traced his finger down the wall of his hole. It was like a window into the earth. What did it all mean? he wondered.

One of his school mates was equally absorbed in the mysteries of the earth and together they pored over his colour illustrated encyclopedias, especially the volume devoted to minerals. However, it was a chance meeting with a retired miner from Broken Hill that really lit the geological spark in young Reg. Beckoning him into the dusty darkness of his tool shed, the old miner pressed three small mineral samples into Reg's hand—a gleaming, metallic-grey piece of silver-lead ore from Broken Hill, a thin slice of shale splattered with vivid green malachite and blue azurite from the Burra Mines, and a colourful chunk of quartz-siderite-chlorite vein material from the local Glen Osmond Mines. The 10-year-old was ecstatic.

At home he laid the rocks out on the garden seat and admired them for hours. He squinted against the points of sunlight sparkling from the shiny metal faces of the silver-lead and marvelled at the brilliant blues and greens in the sliver of drab shale. How wonderful it would be to find some samples of his own.

Burra and Broken Hill were both out in the country, several hundred kilometres away, but the Glen Osmond mine was readily accessible—just beyond Adelaide's eastern city limits. It immediately became Reg's regular weekend haunt. There he found shiny cubes of galena and bevelled, tabular crystals of barites to add to his collection. When his enthusiastic fossicking uncovered an unfamiliar treasure he took it to the museum for identification. At the end of 1929, the museum's honorary curator of minerals, the esteemed Antarctic explorer Sir Douglas Mawson, was away down south on his third trip to Antarctica but this was of no consequence to Reg. He made his way to the university next door and there sought out the geology department's Dr Cecil Madigan, who was happy to answer the young boy's questions.

As Reg's rock-collecting forays extended further from home, his visits to Madigan's rooms became more frequent. On rabbit-hunting trips with his father, Reg would sneak off to inspect rocky hills and outcrops while Pop set the traps. Pop was even taking a stab at gold prospecting to boost the family's depression-era income, and sometimes Reg was allowed to go too. With Great Uncle Ted they scoured all the likely sites in the Adelaide Hills to no avail and then struck out for the Teetulpa goldfield near Yunta on the road to Broken Hill. There, the manager of the outstation showed them an old Aspro bottle full of small gold nuggets and pointed them toward a likely looking 'ironstone' reef that had been discovered while digging a fence-post hole.

For a week they wielded their picks and crowbars to follow the red seam down two metres into the ground. By night they heated the ore, doused it with water to make it crack, and then took turns grinding it to powder in the big, round dolly pot, all the while straining their eyes for any glimmer of gold. Aching muscles, skinned knuckles and one small nugget the size of a shotgun pellet was the result of all their efforts. The rabbit skins Pop had stretched out to dry on fencing-wire frames in the back of the utility would net them more. Reg alone among the trio was pleased with the haul: magnetite crystals, devil's dice, lumps of fool's gold and gypsum crystals to add to the collection gracing the shelves in his bedroom back home.

By now Reg was at secondary school, attending the rather exclusive Adelaide Technical High School—an independent, co-educational, non-fee-paying school with a reputation for academic excellence. The school had been originally created to provide preparatory courses to boost the educational standard of young men planning to study at the School of Mines and Industries. It had quickly proved so popular that entrance exams were instigated to limit numbers. Reg was not a natural classroom student and had missed out when he had first applied, but when the school he was attending in its place dropped geology from its curriculum he gained a transfer into Adelaide Tech.

It was the perfect environment for Reg. In the bluestone, Federation-Gothic Brookman Building that hunched on the corner of Frome Road and North Terrace, along from the university, Art Gallery and Museum, he could formally study geology and technical drawing—both subjects at which he excelled. Downstairs from the classrooms was the School of Mines Museum, housing a fabulous array of minerals from the Broken Hill, Burra and Moonta-Wallaroo mines. In his lunch hour and after school, Reg spent many hours there, handling and marvelling at the samples; familiarising himself with their textures, colours and shapes; effortlessly memorising which minerals came from where and what their common bedfellows were. The curator, Harley Hooper, appreciated Reg's innate affinity with rocks and allowed him to help in tending the collection. Reg soon became his trusted volunteer assistant. When Hooper sailed to London for King George VI's coronation in May 1937, he arranged for 17-year-old Reg to officially take his place as Acting Curator for payment of 2s 6d per week.

The weekends, Reg devoted to mineral collecting in the Adelaide Hills and increasingly to browsing at the Adelaide Public Library. He loved reading about the early geologists and explorers who had pushed through the invisible membranes at the cities' outskirts and embarked heroically into the great unknown. Saturday mornings he regularly spent at the home of retired Adelaide University Professor Walter Howchin. When the reverend professor had resumed taking bible class at the Unley Methodist church, Sunday School had suddenly become much more appealing to Reg.

Howchin was a doyen of South Australian geology. He had written the definitive student textbook, *The Geology of South Australia*, lectured at the university for 18 years, and indefatigably combed the countryside in all seasons wearing a black suit with swallow-tailed coat, stiffly starched collar and black bow tie, along with his cloth cap and leather leggings.

At church one Sunday, he invited Reg to visit him at his home the following Saturday and he would dig out a specimen of fossilized *Archaeocyatha*. Reg needed no second urging. In his rear-garden shed Howchin described to Reg where he had found his first specimens in 1897, and where Reg should search for them himself on Yorke Peninsula. Then the discussion turned to other matters geological: tales of past field trips and Howchin's theories on South Australian glaciers, which had riled some of his more scientifically conservative colleagues. Reg listened avidly. He was entranced with the old man with the white whiskers and small, round spectacles. For Reg he was a grandfather-figure with interests aligned with his own. He treasured their sessions together.

Thus did Reg Sprigg's geological education proceed. It was *ad hoc* but of the finest calibre. Men like Howchin, Madigan and even Mawson tolerated the pesky young Sprigg and his endless questions, realizing that his flouting of early twentieth century Adelaide societal conventions by approaching their learned selves was fuelled by an exuberant inquisitiveness that was impossible to resist.

AT THE BEGINNING of 1937, Reg Sprigg turned 18 and began his first year at university. He was greatly relieved to be there. Gaining his matriculation certificate had required night classes in French and three attempts at passing the exam. For drawing and geology he had achieved credits, even though by then, Adelaide Tech, too, had dropped geology from the curriculum and Reg had had to study independently to sit the exam. Incongruously, he topped the state in geology despite being at a school not teaching the subject.

Geology was just not a fashionable career choice back then. So few was the number of professional geologists in Australia at the time that they knew each other on first name terms. Even the fall-back position of a posting as a secondary school science teacher was in jeopardy as more schools dispensed with geology due to lack of student interest. For many first-year-university science and arts students, geology was merely seen as the 'soft option' for gaining the requisite number of units. Even Reg—rockhound that he was—had no firm aspirations for his post-graduate years. With only two or three university positions, and two at the state Department of Mines, options for geological employment were few. For Reg, back in 1937, a university education was simply fulfillment of an avaricious desire to learn as much about the earth as he possibly could.

At the time, Adelaide University's Department of Geology was the province of Douglas Mawson and Cecil T Madigan. Polar explorers and geologists both, they were the lifeblood of the department for more than twenty years. They were big men in all senses of the word: Madigan, tall, trim and charming, easygoing and an excellent raconteur; Mawson, tall, solid and imposing, titled and reserved. Adelaide University geology

Students Alderman and Thomas accompany Adelaide University Professor Sir Douglas Mawson and lecturer Cecil Madigan on a geological excursion from Mount Mary to Burra, May 1922. *Left to Right:* Mawson, Arthur Alderman, Madigan, Richard Grenfell Thomas and 'The Gums' Station manager Mr Pearse. *(Thomas family collection)*

students were blessed to have two such fine men dedicated to the education of their minds, however, even men of great stature can stoop to petty taunts and arguments when confined to a small department, over which only one could be the boss.

Reg, and other students of the era remember the strained relationship between the two men. Sir Douglas was the professor and Dr Madigan the senior lecturer. Both had risked their lives in Antarctica, both returned home in February 1914 to civic receptions and relieved fiancées. Mawson had survived the remarkable lone trek across the ice after Ninnis and Mertz perished; Madigan had volunteered an extra year on the frozen continent to await Mawson's return. The resolve of both men was astounding. However Mawson also came home to a knighthood and, later, a university promotion. One of his first decisions as the new Chair of Geology was to appoint Madigan his second-in-command. But the gap between professor and lecturer was wide and a natural-born leader like Madigan found it difficult to accept the junior role. Disputes from Antarctica spilled across the years and it was in the halls of academia that Madigan's bitterness festered.

'There was a more or less continuous bickering about how to do things,' Reg remembers. 'Madigan had had [Second World] War service as a colonel and he believed he was a better organiser of men. He often said that Douglas couldn't organise things. "If I'd been down there I would've organised things differently," he'd say.'

'Madigan was terribly in Mawson's shadow,' agrees former student Keith Johns. 'Mawson got all the accolades and he was just in the shadows, and he was bitter about some aspects.'

Johns was on a second-year geology field trip with Madigan in 1945 that ended with a car crash and Madigan in hospital with a broken arm and gashed head: a mishap that unleashed 20 years of frustration in Madigan.

Madigan did not own a car and had no driving licence, and yet felt that it was his right, nevertheless, to use the departmental Dodge utility. Mawson refused to authorise university funds to pay for a licence for Madigan. Madigan wouldn't pay for the licence on principle.

Thus students would drive the vehicle on Madigan's field trips down south of Adelaide to the Fleurieu Peninsula and Kangaroo Island. On this occasion one of the students took a bend in the dirt road too fast and the vehicle overturned.

'Madigan got a broken arm and he had blood running down his neck. We took him to the hospital at Victor Harbour and Madigan really vented his spleen on Mawson,' Johns remembers. 'All the things that he otherwise wouldn't have talked about, I'm sure, but the circumstances were such that he poured out all this poison—all these things that were on his mind about Mawson. And it started really with Madigan not being in possession of his licence.'

Reg relates that when Madigan appeared at the refectory lunch table a few days later with a heavy, turban-like head bandaging, Mawson was hardly sympathetic. 'Madigan's just being showy,' he remarked for all to hear. 'He looks like an Afghan.'

For Reg, it was upsetting to be an unwilling party to the antipathy between two men that he held in such high esteem. He felt it was damned bad luck that two such great figures were virtually competing over the loyalties of students.

'Mawson at times did show unforgivable favouritism, and sometimes as if to pique Madigan's choices. He was very conscious of the fact that Madigan used to say some terrible things about him to the students. And Madigan would very often say them so Mawson—down below—could hear, and this used to embarrass his students.

'Dougie could say some pretty strong things at times and so could Madigan. But I think Madigan was often harder—talking very openly. Mawson didn't do that. He'd say, "silly old man, he doesn't know what he's talking about," but other than that there was no bitter recrimination there. Whereas with Madigan, there was.'

All the students, like Reg, tried to tiptoe down the thin line where the personalities of their two masters clashed. The design of the geology curriculum made it a little easier on them. Mawson took the first year lectures, Madigan the second, and third year subjects were allocated by speciality.

At the end of 1938, when Reg had completed his second year of studies, he was offered a cadetship in the department. In return for free tuition and a stipend of 10 bob a week ($1) he would work as the Professor's assistant. It meant it would take him longer to complete his degree, as fewer subjects could be taken per year in order to fit in departmental duties, but this was irrelevant to Reg. It was a highly sought after position and Mawson had received: 'many applications to fill the position, amongst these none so good as that of RC Sprigg. As Sprigg's main interest in life is geology,' Mawson advised the university Registrar, 'he would be a great acquisition to our department.'

Only one cadet at a time was appointed and Reg's opportunity had come when Lee Parkin, his predecessor, had finished his degree and gained a job as mine geologist with North Broken Hill Ltd. Lee was the first geologist employed by that company and his appointment gave the other senior students hope that a professional geological appointment wasn't such a fantasy after all.

'I started in 1934 as a cadet, doing a five-year degree,' says Lee. 'In my third or fourth year Reg turned up. I was interested in science, but I didn't plan to do geology…unlike Reg. Reg was a geologist from birth almost. I remember him coming in as a schoolboy in short pants, dropping in with a bag of rocks.

'My impression was that Mawson thought he was precocious. And he was, too. Very precocious. He had an utter confidence in himself. Always. And was never inhibited by the niceties of this or that. He'd come into the professor and argue the point about a rock, when he was only in his teens. Mawson found him aggravating, because he'd argue the point, and he'd often turn out to be right! However, Mawson encouraged him because Mawson could see someone who was competent and enthusiastic.'

Like Lee, Reg's first job as the new geology cadet was to clean out the crypt beneath the Prince of Wales lecture theatre. In this musty, dark, windowless space the cadet would sort out mountains of minerals that

Mawson had squirreled away for further study. Many were labelled with sample site descriptions such as, 'near lunch camp, Tuesday,' or 'near horse and cart, Wednesday,' which only Mawson, himself—perhaps—could make sense of. There were also bags and bags of radioactive ore from the Radium Hill mine, amongst which the students innocently shuffled about for hours at a time. Reg regarded cleaning up the crypt as 'disappearing into its darkness, rattling amongst collections of sledges and making sure to be seen or heard thus by the professor,' who rarely went in there himself.

Another of the cadet's duties was to mark attendance at lectures. In those days it was not a matter of casually turning up if you felt like it. Attendance was compulsory. All the seats sloping up to the rear of the auditorium were numbered and each student was to sit in his or her assigned seat. The cadet would then carefully record the vacant seats for the professor's attention.

'The way they'd beat you, was to shift seats while you were counting,' laughs Lee. 'No one took it very seriously; I don't think Mawson checked it. But we had to mark the roll.'

It was protocol only for first year lectures, when over 100 students filled the theatre. By second year the group had shrunk to no more than 20 and all the faces were familiar.

With the roll complete, Mawson would stride in, in his black, academic robe and begin the lecture in his characteristic clipped tones—every 'p' popping and every 'c' and 't' in 'Antarctic' clicking. A sharp tap on the floor with his wooden pointer would signal the picture slide operator for the first glass slide. To his face the students always addressed the professor as 'Sir,' but behind his back the 57-year-old was affectionately 'the Old Man' or 'DM.'

Other tasks allocated to the cadet were cataloguing and rearranging material in the departmental museum, where rocks, minerals and fossils reposed in glass-topped cabinets; reproducing lecture notes by the ancient cyclostyle machine; ensuring students' laboratory specimen sets were complete and pilfered samples replaced; mastering the art of cutting transparently thin sections of rock for microscopic examination; and arranging details of field excursions. Whether in the department or out in the field, the cadet was the general 'dogsbody.' It was both a privilege and a chore.

For Reg, the major benefit of being a cadet was being invited to the geology common room—nicknamed the boat room—to enjoy tea breaks with the senior staff. There, sitting beneath assorted equipment from Mawson's various Antarctic expeditions stored in the rafters above, Mawson and Madigan would engage in discussions and arguments to fascinate their companions.

'You were never short of something to talk about when Mawson was around,' agrees Keith Johns. 'He always dominated the conversation really.' And sometimes the topic moved south, to the Pole. One of Mawson's favourite anecdotes was to describe killer whales catching seals on an icefloe.

'The orca's head emerges vertically at the edge of the floe to judge where the seal is laying,' he'd say, peeking the head of his spoon slowly over the edge of the table. 'The whale swims under the floe to the precise point under the seal…and comes up, Bump! to break the ice,' at which point Mawson would crack his fist on the underside of the table, startling his listeners into seal-like empathy.

It was on field trips, though, where the students generally got to see a more relaxed Mawson. The hiking excursions were serious learning opportunities—'It is no good just collecting minerals for the sake of collecting,' he'd say. 'You've got to find out why it is here in the first place. Use your eyes. Don't just walk along like a mob of emus. Every pebble tells a story, if only you will take the trouble to interpret it.' But it seemed to be mealtimes—when Mawson took charge of the cooking pot and recreated his polar camping days—that he most enjoyed. Emerging early from his tent in his trusty Jaeger wool pyjamas he would prepare breakfasts of porridge and eggs for his students. Dinner was usually a meat and vegetable stew prepared from a 'feeder' stock from the day before. He would routinely retire the fat for another meal 'down the line,' and dish out the watery intermediate layer as energy-giving soup 'like in Antarctica.' Survival habits were hard to shrug off, it seemed.

'Mawson was not much given to "fancy foods,"' according to Richard Grenfell Thomas, one of Mawson's students in the early 1920s. 'He usually preferred things he had become used to on his Antarctic travels. These, in our opinion, were seldom suited to summer temperatures at Boolcoomatta! Chocolate slab, sardines in oil, tea heavily sweetened with condensed milk and a peculiar form of "everlasting" bread, comprised the main diet. We used to measure the amount of poured condensed milk by the succession of ants that came out on the thin stream of milk!'

There were other Mawson peculiarities, too, that stuck in Thomas's mind.

'DM was a bit of a terror as far as fences were concerned. If we encountered a fence without an obvious gate he would consider whether we could drive over the fence by having some of us stand on the fence to flatten it! If that was impossible he would, on occasions, cut the wire in the fence with geological hammers and then we would try to mend the fence afterwards with spare wire, which we always carried.'

The 'vehicle' that Mawson used in Thomas's time was an old spring cart and a horse called Polly that he hired from Harry Ding in Olary.

'We used to get out of the Broken Hill express at Olary at about 4.30 am and try and keep warm until daylight around the peppercorn trees at the small railway station at Olary,' Thomas recalled in a letter to Ding. 'On the 1921 occasion I remember very well that as soon as it was "decently daylight" we all went across to the store your people ran at Olary. After some delay a young girl, who must have been your sister, came into the shop and promptly retired to the back part calling out, "Mum, come quickly, the shop is full of men!"' Thomas found it amusing that he and his two fellow students, Arthur Alderman and Clarrie Piper—each about 20 years old—had the distinction of being classified as 'men.'

Mawson taught them how to measure the distance they traversed in their geological mapping by tying a handkerchief to a spoke of the cartwheel and counting the number of revolutions of the handkerchief. They could then determine the mileage by calculating the circumference of the wheel and multiplying it by the number of revolutions per minute.

'Another of DM's frequent peculiarities was that he often neglected to button up the fly of his pants!' Thomas recalls. 'I remember once at Quorn, he had taken time off to call on the Mayor. After returning to us students after the formal visit he discovered his fly was open from top to bottom, but he didn't seem much embarrassed. Then again, once at Adelaide University DM had been attending some sort of levee or something at Government House. He was attired in some sort of formal court dress and returned to us students with his trousers fly fully open, as usual. Several of us could not refrain from laughing in an embarrassed way, at this spectacle, whereupon DM, who seemed mystified at the cause of our mirth, became confused and sat down on a valuable brass template gadget used to draw stereographic projections of crystals. The instrument promptly broke into several pieces which added greatly to the general confusion!'

Thomas's student trips with Mawson were before Reg's time, although he and RGT, as Thomas was frequently referred to, became firm friends in latter years.

Many of Reg's recollections of Mawson in the field revolve around a private visit to the northern Flinders Ranges in late 1939, when Reg was nearing the end of his third year of studies. It was their second Flinders trip together. Unlike their visit the year before, when Reg had been one of a group of six second-year students, this was to be just him and Mawson, with 'Smiler' Greenwood joining them at Arkaroola. Reg felt exceedingly lucky to be included.

It was September 1939, when they were packing up to leave. The papers were full of Hitler's hostile taunts to the Empire. A second world war—once imponderable—seemed inevitable. Mawson's immediate concern was for his in-laws in Holland, so close to the heavy heel of the German jackboot. At one point he had considered abandoning the trip, but with the new portable wireless his wife had bought him he could travel out yonder and still keep abreast of world affairs. The new technology was marvellous, Mawson agreed, as long as it was restricted to the appropriate purpose.

To begin with, he had vehemently opposed the contraption. When he had returned from work one evening and found a brand new wireless

Mawson packing geological specimens, Encounter Bay geology camp, August 1922. *(Thomas family collection)*

cabinet in the lounge room of his Brighton home he had turned on Paquita in a fury. 'If that thing's not out of the house by tomorrow afternoon when I return from University—I'll put an axe through it!' he roared. Such mindless entertainment would detract from family living, he admonished, and was particularly bad for the children. Paquita did not always bow to her husband but this time she didn't argue and returned the machine to the store forthwith.

When world politics began to dominate the family dinner-table conversation, however, and Mawson was desperate for timely news, Paquita realised the time was right for a second attempt. What about one of the new portable radios, the salesman suggested, he can have it with him at all

EARLY DAYS

times and at least it will be there for the family to share at night. To this, Mawson acquiesced, and the three-valve portable radio became his constant companion.

'God, how all of us in the Geology School came to detest the bloody thing,' Reg remembers. 'A flick of the lid and it switched on. By closing the lid it shut up. Never a news session or an "on the hour" report opportunity passed without it blaring forth the latest on Europe, thus killing any hope of intelligent conversation. Previously, prolonged tea breaks had always been justified amongst departmental members as presenting opportunities for learned discussions. More often, it was a case of competing anecdotes between Mawson and Madigan—if not about Antarctica, then about developments in Central Australia. All this was now a thing of the past. A sense of defeat descended on the department. Fiery exchanges between the two Heads of school now centred on Nazi intentions and the all-too-slow arms build-up in a reluctant West.'

Naturally the radio was amongst the items to be packed on board the Dodge buckboard for Reg and Mawson's excursion to the Flinders. Reg attempted to stow everything safely as Mawson passed a parade of duffle bags and camping equipment up to him, perched as he was on the back of the utility tray. Apparently Reg's arrangement was not to the professor's liking... 'Oh for God's sake, Sprigg. Get down off there and let me do it. You couldn't manage a night cart!' Mawson said, as he exchanged places with Reg and set about reorganising the load.

'Pass me the tucker box,' he ordered Reg. 'Not that way, turn the thing around. Now the camp oven. Did you have to drop the cursed lid! Where are the swags? Well don't just stand there like a dummy...get the things.'

Reg was anxious to please, but it seemed he couldn't do a thing right. They hadn't even departed the university. Was the whole trip going to be like this? he wondered.

'What a shambles here, Sprigg. You've given me everything in the wrong order. You'll never make a field man. Now, pass me the plate camera.'

It was the last article to be stowed. The Old Man's precious Zeiss camera that had captured geological formations from around the country and icy

21

images from the end of the earth. He'd bought it in 1906 for the lavish sum of £9 12s and it had been an essential part of his field kit for the past 30 years.

Reg hoisted the camera, encased in its heavy leather box, high above his head towards the professor's impatient, outstretched hands. He felt the weight of the camera evaporate, as Mawson snatched at it…followed by a sickening crash. The camera had fallen more than two metres to the ground and there was Mawson towering above with just a worn leather strap in his grasp. The handle had pulled free from the case and the plate glass camera was now a boxful of tinkling remains.

'That camera has been with me just about everywhere Sprigg,' Mawson castigated him unfairly. 'It has catapulted from sledges, hit the deck, fallen overboard, been left on icefloes and suffered every possible indignity, and now you have it for two seconds and you have to drop it! You've buggered the whole expedition.'

So began the excursion that Reg had been anticipating so eagerly. It was a long and silent drive to their beds that night.

The next day they made Copley. The wound from the lost camera was beginning to scab over and Mawson was in good spirits as their store and post office chores were over, with enough time to catch the 3.30 pm radio news. It was the international news session—the one he most hated to miss. Parking the car near the petrol bowser for refuelling he jumped out of the driver's seat and retrieved the portable radio from the load, setting it on the right-side, front mudguard. While Reg made notes of expenditure, mileage and so forth, Mawson twiddled the radio dials in increasing frustration. He then ran out 20 metres of aerial wire, put some water on the ground and inserted an earth wire into the dampness to improve the reception. But his efforts were in vain—only static and an indecipherable soundtrack of electronic beeps came forth from the instrument.

He slammed the lid down in disgust.

'Move over Sprigg, you can drive the next section,' he commanded, shoving himself into the passenger seat in irritation.

Determined to make a good fist of it on the slippery, mud road, Reg focused on the wheel tracks with single-minded concentration. He failed to see the radio still balanced on the mudguard until it was too late. Again, there was a splintering crash. This time it was the coveted radio…tumbling along behind them…disintegrating into bits and pieces.

They gathered the broken remains in a stony silence: Reg too afraid to speak; Mawson too angry. Reg poked the remnants into a corner of the load and quickly took his seat. Mawson sank into the passenger seat accompanied by one final, high-noted, goading crunch. His glasses. The Mawson spectacles crushed in a concluding calamity. 'For two lousy pins, Sprigg, I could murder you,' he seethed.

What a trip it was turning out to be!

Charniodiscus arboreus

CHAPTER 2

WARTIME URANIUM SEARCH

WHEN REG AND Mawson returned to Adelaide, Australia was at war. Like most of his male friends, Reg felt it his duty to sign up. He applied to the Air Force but was diverted to the Royal Australian Engineers instead, and reported for duty at Adelaide's Fort Largs in January 1940. His career in the armed services didn't last long, however. Seven months later—following Mawson's surreptitious intervention—he was back at the university to complete his Bachelor of Science in zoology and an honours degree in geology by the end of 1941.

With his student days now behind him, Reg tried once again to join the Air Force, but by then the Manpower Priorities Board was managing employment across the nation and allocating men and women among the

21 000 essential industries and services declared 'protected undertakings.' The omnipotent agency vetoed Reg's desire to join the military. His scientific training could be put to better service within Australia, it decreed, firstly supervising the manufacture of cordite and other explosives at the Munitions Factory at Penfield, in South Australia, and then conducting soil surveys for the Council for Scientific and Industrial Research (CSIR, which later became the Commonwealth Scientific and Industrial Research Organisation, CSIRO) in preparation for a postwar 'Soldier Settlement' building boom.

It was while working for CSIR in 1944 that he was summoned by his boss and told that Ben Dickinson, government geologist and new director of the Department of Mines, wished to see him concerning a 'special project.'

Reg knew Ben well. He was a fellow member of the university Geology Club and the two young men—only seven years apart in age—shared a passion for geological mapping. Dickinson now appealed to this joint interest of theirs in asking Reg to join the department's Geological Survey division on a top-secret uranium survey. The project would be conducted on a strictly 'need to know' basis and the motivation for the survey was not revealed to Reg. The war would be over before he would learn that the state's two uranium deposits—the only ones in Australia and two of only a handful known around the world at the time (the others being in Arctic Canada, the Belgian Congo and Czechoslovakia)—were being checked at the request of the British government for supplies of uranium for production of an atomic bomb.

Even Dickinson, himself, may not have known the true nature of the request. When the former state government geologist, Dr L Keith Ward was asked by the Commonwealth Department of Supply and Shipping in November 1942 to reserve South Australia's known 'uranium-bearing areas' for the Crown, he was told only that 'this action is being taken because of the probability that the element uranium may assume considerable importance as a source of power.' Nothing to do with weapons was mentioned, although Ward was astute enough to suspect that the request was related to military requirements.

25

He duly excluded the old radium and uranium ore workings at Radium Hill and Mount Painter from the State Mining Act so that private interests could not acquire mining rights over them, and waited for further advice.

By the 7 May 1944, when Prime Minister John Curtin requested 'immediate steps taken to investigate the possibilities of the production of uranium from Australian ore deposits,' Ward had retired and Dickinson was in the chair. It was his job to assemble as much information and as many men as he could muster and report back to the government within six months on South Australia's uranium potential. It was a tremendous undertaking for a young, 32-year-old man who had been in charge of the department for only three months.

Reg Sprigg was one of the first outsiders Dickinson approached to boost the department's slim ranks. When Reg heard Dickinson's appeal in June 1944 he couldn't sign on quickly enough. Uranium minerals didn't scare him. He had lived with boxes of the stuff under his bed for years—rich specimens of bright green torbernite and canary-yellow autunite pilfered in his school days from a School of Mines rubbish dump. He could only wonder, why such interest now, in deposits that had been discovered three decades ago.

It was not his job to ask questions, however. His job was to load up the Chev buckboard and departmental caravan with his store issue of supplies including miner's pick, Dolly pot, shovel, axe, small tent, stretcher, billies, hurricane lamps, butcher's knife and frying pan; along with provisions such as 6 dozen eggs, 3 pounds of barley, 2 cabbages, 56 pounds of potatoes, 2 pounds of swedes, 2 pounds of rhubarb, 3 tins of powdered milk, 12 tins of condensed milk and 6 pounds of flour, among other things—and head east for Radium Hill, an abandoned mine site about 110 kilometres south-west of Broken Hill. With his secondment into the Department of Mines he felt he would finally be doing something important to help the war effort. And what fascinating work it would be, too. Tracing the outline of the uraniferous lode, exploring the old mine shafts—what more could a budding young geologist wish for! He couldn't wait to get started.

Within a matter of days he was on his way.

Radium Hill—more of a pimple than a hill—was the site of Australia's first uranium mine. When prospector Arthur Smith, searching for tin, had found a strange rock there in March 1906 he had sent it to the Department of Mines for identification. They forwarded it to Adelaide University for the attention of the newly appointed 23-year-old mineralogy and petrology lecturer, Mr Douglas Mawson. Mawson had an interest in radioactive minerals and had readily identified the yellow dusting of carnotite on the dense, black rock. He subsequently named the heavy, black, primary uranium-bearing mineral, 'davidite' in honour of his former professor at the University of Sydney, Edgeworth David.

Smith proposed to Mawson that they form a partnership to exploit the find: In return for funding and conducting the geological investigation, Mawson would be entitled to half of the profits from the radium extracted from the radioactive ore. At that time the newly discovered radium—a disintegration product of uranium—was the prize. Uranium, itself, was virtually worthless, finding application mostly in giving an orangey, fluorescent tinge to glassware and glazed tiles, and in the making of luminous paints for watch dials.

Mawson soon found that scratching davidite out of the earth as a source for radium, was no way to make money and the lease expired on the claim with neither Mawson nor Smith finding their fortunes. A couple of later mining attempts were made by various syndicates, but despite pure radium bromide fetching a phenomenal price of up to £17/mg in the early 1920s for use in medical radiotherapy, the extraction process for the Radium Hill ore was so complex as to make it economically unviable. The mine was finally abandoned in 1931.

When Reg arrived on site in late June 1944 there was very little to see. A couple of concrete slabs, some low mounds of waste rock, torn bags of ore concentrate strewn about and several rough, yawning tunnels into the ground was the extent of it. The country was in drought—dry and desolate. The knee-high thickets of bluebush were drained and dusty; a few faded mulga trees clung to the blowing dirt. Here, in the back blocks of Oulnina Station all was silent, save for the buzzing of the bush flies.

For Reg, though, it was invigorating.

He unhitched the caravan and set about making camp. This would be home for the next few weeks. Tomorrow he would start work.

His task was to open up the old mine workings, geologically map them and assess the uranium potential. 'Make a detailed survey of the workings and map the geology carefully,' Dickinson advised him by mail. 'In the latter respect, in particular, note the width of the lode formation from point to point, and any marked changes in strike and dip.'

There were five original shafts on the 'hill' and Reg explored each of them in turn, clambering about by the glow of a carbide light; measuring tape, notebook and geological pick fastened to his belt. He measured the lodes, shafts, drives and stopes, and recorded all the surface geology. Every couple of metres he hammered out a rock sample and secured it in a calico bag.

He worked largely alone, dangling on ropes in shafts up to 50 metres deep; slithering through tunnels, narrow and crumbly; dragging himself through a 50 centimetre-wide channel created by a previous rock fall; live snakes; dead lizards; cramped conditions—it was nerve-wracking, dangerous, difficult work. But he loved it! The geology was fascinating and that was all that mattered. He returned to camp at sunset, filthy and tired but revelling in the wonderful opportunity that had come his way. At night he worked by Tilley lamp drawing up plans and labelling his rock specimens for analysis in Adelaide. The raging conflict encircling the world seemed hard to imagine from his spot of splendid isolation.

After two months at Radium Hill—and some assistance in the latter weeks from the men of Zinc Corporation Ltd, which company had previously held the mining lease for a short while—Reg's work at the old radium mine was complete. Although he'd concluded that Radium Hill still held definite potential as a uranium mine, he was not to know that the urgency of the wartime uranium search virtually excluded the site, as there was still no convenient way to extract the uranium from the complex Radium Hill ore. He packed up camp resigning himself to a return to CSIR and his soil survey work—which now seemed even less engaging, by comparison.

However, when he arrived back in Adelaide there was good news. His secondment had been extended and he was now to join the geological team investigating the uranium deposit at Mount Painter—the state's other uranium field—about 300 kilometres north-west of Radium Hill, in the northern Flinders Ranges. The Mount Painter uranium ore was more amenable to simple methods of recovery and all attention now focused on option number two. The remote arid rangeland, 650 kilometres north of Adelaide, was a flurry of activity.

Like the discovery of the Radium Hill prospect, Mawson had had a hand, too, in the development of the Mount Painter uranium field. In 1910, William Bentley Greenwood, a local pastoralist and part-time prospector had called on him with crystals of an unknown, shiny, grass-green mineral that the government assayer at the School of Mines had identified as iron phosphate 'of no commercial value.' Greenwood was dissatisfied. He had a hunch that there was more to it and took his 'stone with the green crystals' to the geology department at Adelaide University. He had previously sent Mawson samples of corundum—raw sapphires and oriental rubies he had found in 1905—and trusted the young lecturer's judgment.

Mawson gave the green material careful inspection, recalling the advice given to him by radium expert Marie Curie on his recent visit to Paris. If you ever come across a bright-green or -yellow mineral that you cannot identify, she had told him, suspect the new element, uranium—the mother of wonder element, radium. She had shown him a range of radium-bearing minerals and purportedly given him a gold-leaf electroscope with which to detect radioactivity. Mawson now fetched one of these instruments. It had a brass plate at the top and a hinged pair of gold leaves hanging from a rod. An electric charge created by rubbing a silk handkerchief over an ebonite or wax rod and transferred to the plate, caused the gold leaves to repel each other and spring apart, rather like dry strands of hair combed with a plastic comb. If a source of radiation was held near the spreading leaves they would lose their opposing electric charges and collapse back together.

Greenwood watched as Mawson brought his sample towards the gold leaves hovering in mid-air and saw them promptly flop together. It was radioactive all right. The green crystals were later proven to be radium-rich torbernite.

Old 'WB,' as Greenwood was known, was an inveterate fossicker. To make a living, he managed various sheep stations in the vicinity, but any spare time he devoted to prospecting deep into the granite hills of Mount Painter country, often in company with his son Gordon, known as 'Smiler.' The Government employed him for a while as an official part-time prospector, hoping that the rugged hills would yield copper in similar quantities to those found at Burra, Kapunda and Wallaroo in the mid-1800s, or commercially viable amounts of lead, zinc, or silver like the fantastic deposits discovered at Broken Hill. While overseeing Beltana Station, WB had discovered the silver and lead outcrop at Ediacara which had spurred a short-lived silver rush in 1888. He seemed to have a nose for such things.

In 1910, WB was caretaker of Umberatana Station, on the plains 20 kilometres west of Mount Painter. He often ventured into the rugged hills out east, hunting for goats to supplement his meat supply, at the same time always on alert for items of geological interest. This was feral country in all senses of the word; wild and inaccessible and over-run with vermin. Each of the surrounding stations had erected fences across their border with the Mount Painter environs, preventing their sheep from straying into the rocky hills and keeping out the dingoes, rabbits, donkeys, camels and goats that multiplied unchecked.

It was during one of these trips that WB had returned to a former campsite, 'Echo Camp,' and headed northwards along a dry creek, retracing his steps of a sortie made a decade earlier. A large arrow he had composed in 1898 with white quartz rocks, still pointed the way. Father and son worked their way up a steep slope above the creek and it was here that Smiler had found the splashes of lime-green, clinging to the rock like flaky, green fish scales. When Mawson accompanied them later in the year, they located the torbernite lode, high on what was to be called 'Radium Ridge,' a prominent ridge west of Mount Painter. To Mawson, it was a striking feature of the surrounding country: A serrated ironstone ridge rising as much as 1000

An early bulldozer forges the first track through East Painter gorge during the Mount Painter wartime uranium survey, late 1944.

feet above the adjacent creek beds. 'As far as I am aware,' he wrote in his field notebook, 'this is the most extensive uraniferous lode formation in the world.'

Mawson and Greenwood formed the Radium Extraction Company of South Australia and old WB led a prospecting party of five men equipped with camels and equipment to begin primitive mining in December 1910. Like the fitful attempts at Radium Hill, operations continued sporadically until the early 1930s but no fortunes were to be made at Mount Painter either. These radioactive minerals were easier metallurgically to extract from the parent rock but they were dispersed over a large, ill-defined area. In contrast to Radium Hill, there was no contained body of high-grade ore, and rich surface veins soon faded out at shallow depth. Too, there was no nearby railway siding. The rock had to be carted by camel and then bullock-wagon for many miles to the rail-head at Copley. It was said that the continual rubbing and jostling in the camel bags and boxes ground the rocks away, with the precious radioactive fragments falling from cracks and splits in the seams; trampled into oblivion on the dusty trail behind them.

Reg arrived at the Mount Painter site in the middle of October 1944, and joined the working party. Investigations had begun in June and by now, 70 men were working at the field. Despite the shortage of manpower dictated by the war, mines chief Dickinson had every Commonwealth department and the CSIR at his disposal and was able to draw miners, prospectors, engineers, surveyors and geologists from wherever he could find them.

New underground drives and exploration adits were being cut high into the hillsides where the deposits mostly occurred. A sled was improvised to haul mining machinery to the various working levels, but much equipment had to be physically manhandled up the steep slopes. A dirt aerodrome was being constructed on the plains at Balcanoona, 40 kilometres south, to fly in emergency supplies and key personnel. A new road from there to the mine site was also being formed, replacing the Greenwoods' old camel pads. Smiler and his younger brother, Bentley, who by now had established Arkaroola Station—enclosing Mount Painter within its borders—shared local information and provided assistance in the form of camel strings for carting while the new road took shape.

A camp village had sprung up near to the most productive 'Number Six' workings above Radium Creek. Camp water came from a bore four miles away, however, drinking water had to be carted 32 kilometres from neighbouring Wooltana Station. Soon a second camp complete with mess hall and galvanized iron 'camel cubicles' instead of tents for worker accommodation, was established at East Mount Painter when even richer concentrations of torbernite were found there. The bulldozer provided by the American Army—remarkable for being the first large dozer that Reg and many of the others had seen—was put into further service creating a 27 kilometre-long road through the boulder-strewn East Painter gorge to connect the two camps. It took 68 days to complete…more than two days for every impenetrable kilometre.

Reg was generally living in less commodious surrounds out in the bush. In company with assistant, Hector Ward, his job was to survey and geologically map the area west of Mount Painter and search for possible new uranium deposits. Another pair was operating on the eastern side and between them they were expected to eventually cover 220 square kilometres.

Reg and Hec lived a nomadic life on the western reaches, roaming the Arkaroola-Mount Painter ranges that Reg found so inspiring. For him, it was a return to his undergraduate days and he thought often of Mawson and the respect his professor had for this dry, rocky country. He and Hec careened about in Reg's narrow-tyred, two-wheel drive Chev buckboard and slept under the stars on primitive straw mattresses. After breakfasting on bacon, eggs and tea the two would roam on foot—setting off in different directions and keeping in touch via spinifex-fire signal flares at midday. After hiking from dawn to dusk on the sustenance of a couple of apples or oranges, they would reconvene at camp for a dinner of meat and potatoes, which had been stewing in the camp oven throughout the day. It was enough to fuel their relentless walking, but only just. After two months of daily 10-kilometre hikes, Reg was suntanned and wiry, and 8-kilograms lighter.

Often his diary entry for the day finished with, 'Reached home absolutely buggered, in dark,' or 'After quick tea, fell into bed utterly ruined.' One day in late November, Hec left his waterbottle behind and they had to share Reg's: 'one pint each for 11-hour hot day trek over hills!' he recorded. A few days later an 11-mile hike in the heat almost proved too much. With insufficient water Reg almost blacked out and his tongue stuck to the roof of his mouth. 'Started home practically out of water. Hot, sultry day and could only climb coll before Radium ridge in short bursts. Returned nearly caved in at 7.00 pm just before dark. Both were so nearly finished that we resolved that no matter what hurry is needed we can't make dangerous sorties like that.'

The car, too, was rather the worse for wear by the end, for it had been urged to go places that were never meant to be traversed in a motor vehicle. In fact it was not in the best condition to start with, having been rescued from the Engineering and Water Supply departmental scrap heap for Reg's expedition to Radium Hill. Still, with prayerful coaxing, and Hec tossing rocks aside to clear the way, the Chev made remarkable progress into steep gullies and along boulder-strewn dry watercourses.

ROCK STAR

It was one such occasion, when they ventured into the mouth of the great Yudnamutana Gorge—slithering down a rocky, 100-metre-high slope to the valley floor—that nearly defeated both men and machine. There seemed no way back. But as they toiled on, taking hours to proceed a fraction of a kilometre, the way forward became equally impossible. Eventually there was no choice but to attempt a retreat.

By now the car was jammed at the bottom of a narrow, dry waterfall that ended in a confusion of boulders. The two men stacked rocks against the face of the slope behind them to create a better ramp before Reg sought to reverse the vehicle to the top. Instead it lurched sideways and wedged itself tight. They tried jacking it up and packing in more boulders but the

Bentley Greenwood in his 1926 Maxwell Dodge, negotiates the scrub and boulders of Wywhyana Creek, Arkaroola Station, August 1939.

car refused to budge. Reg dreaded the thought of having to abandon the vehicle and explain the scenario to his Chief. There had to be a solution to their predicament. Hec, however, was dubious, to say the least, when Reg announced that the answer was to hack off the trapped mudguard and running board with an axe! Three well-aimed swings and the metal was loosened. With Reg back behind the wheel and Hec shoving from in front, the car lurched away backwards with a god-awful screech.

At the top of the fall, jubilation quickly turned to horror as they surveyed the crippled car; and then to trepidation as they imagined the reaction yet-to-come, back at the depot. For Reg, a vehicle would always be just a means of transport, he had no interest in makes, models and mileage. But he knew that there were others who felt much more empathy towards their vehicles and he knew enough to know that a buckboard arriving back at the depot sans mudguard and running board would not be greeted favourably. He was right, and was berated for the mutilation when he arrived back in Adelaide after his two-month tour of duty. Still, to Reg's mind the old Chev hadn't done a bad job, even if she did return slightly incomplete. Surely his handiwork didn't compare to the driving abilities of the surveyor who had broken twelve utility springs during a mere three-week stint at Mount Painter!

By CHRISTMAS THAT year, Reg was back in Adelaide devoting the holiday break to report writing on his uranium searches. His western survey had revealed no new deposits but the rich finds at East Painter held promise of further discoveries. Operations in the northern Flinders Ranges were in full swing. Premier Playford and the Mines Chief were said to be full of admiration for the workmen who had toiled long hours on Saturdays, Sundays and holidays to maintain the schedule.

'The necessity of the uranium work to be completed in the shortest possible time makes demands which require us to make sacrifices for the fulfillment of the purpose which lies behind it,' Dickinson wrote to Reg in reply to his enquiry about the five weeks vacation leave he had accumulated at CSIR. For Reg was a young newlywed, and his wife was growing dissatisfied at his continual absences from home.

'Sorry to pass on my personal worries,' Reg had written before his return from Mount Painter, 'but I feel that I have certain obligations to my wife besides my personal feelings on the matter. I have not had more than two weeks total holiday since marriage two years ago and that includes a honeymoon!'

It seemed churlish to complain about holidays when men his age were dying in combat overseas but his wife yearned for him to be at home. He felt he had to ask.

It was a wartime Christmas Eve in 1942 when 23-year-old Reg had married 19-year-old Patricia Day. Since then he had been stationed in Tasmania, Kangaroo Island and the south-east of the state, among other places; rarely at a post handy to their home in Adelaide. Now it was Radium Hill and Mount Painter. His assignments were becoming increasingly remote and the separations were long and frequent.

But in the end, the dedication of Reg and the entire team of workers came to nought. In February 1945, the UK Chancellor of the Exchequer informed Prime Minister John Curtin that Australia's predicted yield of 20 tons of Mount Painter uranium trioxide over a period of 12 months, 'was not now of interest to those concerned.' The operation was abruptly terminated by the Commonwealth Government and all the plant and equipment—so assiduously assembled on far desert hilltops—was trucked out and disposed of. The men dispersed back to their pre-Mount Painter occupations. That was that.

Six months later, on 6 August, the true motivation for the Allied uranium search programme was revealed in all its horror. Upon learning of the devastation visited upon Japan, Reg felt the weight of an irrevocable step in the capacity for human self-destruction. He stared at his photographs of Mount Painter—an untouched sentinel of the ranges for 300 million years—and shook his head in dismay.

'The utilisation of energy released from the splitting of the atom opens up a completely new epoch in human history,' he wrote in his notes. 'It is obvious that what was only recently regarded as fantastic has now become possible and inevitable. Man in this new conquest of his environment enters a period in which possibilities of human advancement exist side by side with the gravest and most terrible threats to all civilisation.'

That no Australian uranium had been the agent of such destruction was small consolation.

The urgency of the wartime uranium search was over. However, it was now clear that this mineral with the 'energy of a thousand suns' was a marketable new commodity, and what's more, South Australia still housed almost half of the world's known deposits. South Australia's premier, Tom Playford, quickly realised the potential of atomic power and pressed his mines department to continue its assessments of mining potential. Dickinson seized the opportunity to reorganize his geological survey division and permanently appoint the small team of experienced young geologists the Mount Painter project had bestowed upon him. He knew that there had been few geology graduates during the war, and postwar development of Australia's mineral resources would be severely hampered by a shortage of suitable men. Reg was made Assistant Government Geologist in charge of regional geological work. It was one of four such positions immediately subordinate to the Chief.

For the new 25-year-old public servant it was a dream come true. Fancy being paid to roam around the state examining hills, troughs and rocky outcrops; advising landholders on the best site to drill a water bore; picking over old mines to see if there were any riches still left in the ground; following ridgelines and cliff-faces and figuring out the basic formation of the earth's surface. It was what he had been doing for the love of it in his spare time for years. Now his hobby had become his career.

Reg's main activity for the remainder of 1945 and much of 1946 was, in fact, selecting sites for water bores to extend South Australia's agricultural boundaries. However, Radium Hill always hovered on the periphery. He made several trips back to the abandoned mine site to finalise his investigation and finish his report in the more relaxed postwar atmosphere. Once his specific duties were complete he always took the opportunity to roam out from the mine, exploring the region between Olary and Broken Hill at the eastern border of the state. This was traditionally Mawson's territory and he felt enormously privileged to be delving amongst the remnants of ancient glaciation that had so fascinated the Antarctic explorer.

When Reg had told his former professor that his appointment to the Department of Mines had become permanent, Mawson was unimpressed. He was loath to see Reg devoting all his energies to the State Geological Survey. Despite taking the opportunity to 'knock him down a peg or two' at times, the professor felt a genuine fondness for his young protégé, whose aptitude for geology was obvious. His talent would be wasted in the Survey, Mawson believed, which, according to him, was populated by field geologists with no imagination whose brains were in their boots. Like his ongoing acrimony with Madigan, his antipathy with the Department of Mines had long roots, but the two institutions, too, were developing very different characters. The cloistered, insular, staid, somewhat incestuous environment of Adelaide University academia stood in contrast to the free-wheeling, young, energetic aura of the State Geological Survey, with Dickinson often luring away the university's best geology graduates.

Reg, however, maintained amiable relations with his former professor and tried to keep a foot in the academic door by devoting all of his spare time to geological mapping. If uranium was the new mistress, then mapping was his true love. He had an innate ability for structural geology; visualising the way the layers of the earth's crust had formed and how they had folded, warped and slipped over millions of years.

In 1944, amidst all the haste and secrecy of the wartime uranium surveys he had still found time to complete his Master of Science degree, analysing the physical geography and development of the ranges fringing Adelaide's eastern suburban boundary. 'The really new contribution to knowledge is contained in his geological map,' Mawson wrote in his assessment of Reg's work. 'There is embodied in it a great deal of patient work and ability to decipher involved geological structures in the field.'

Now, as work assignments took him all over the state he used every opportunity to record the stratigraphy and structural geology of the areas he traversed—documenting what kinds of rocks were exposed at different sites; how they were arranged in layers that folded in gentle curves across the countryside or abruptly disappeared into the earth at steep angles; and where they reappeared in the same formation often many kilometres away. Each kind of rock or substrate was allocated a different symbol or colour and recorded on his regional map, so that, eventually the geological structure of an area could be gleaned at a glance. He thought of geological mapping—as this process was called—as akin to working on a giant three-dimensional jigsaw puzzle, with many of the pieces obscured or missing. It required imagination and the ability to think in three dimensions to decipher how the planet's surface had been pressured and molded.

He made use of every railway cutting, abandoned mine, bare rocky outcrop and naturally weathered hillside to view a protruding vertical slice of the earth's crust—in this way 'connecting the dots' and building up a picture of the overall geological composition of the northern Mount Lofty and Flinders Ranges and an understanding of how they had formed.

IN MARCH 1946, Reg was sent on assignment to a remote area 600 kilometres north of Adelaide. It was a rare occasion where his wife was able to accompany him and they were revelling in the chance to go hiking and camping together. Time together had become increasingly precious as Reg spent more and more of his time in the field.

His first job on this trip was to proceed to an abandoned silver and lead mine and help colleague Ted Broadhurst map the surrounds. The Ediacara mine was in low hills on an arid plain bordering the dry Lake Torrens saltpan. The rusty-looking sandstone of the range stood in stark contrast to the dazzling whiteness of the salt-lake 13 kilometres to the west. In between there was a panorama of red sand dunes that glowed like embers in the evening light.

The couple made camp a short distance away within a corridor of grand old red gums lining Sundown Creek. The gravel in the dry creek-bed was bright purple—mainly slate and quartzite, Reg noted with interest. Pat shouldered the load of establishing camp while Reg limped along after

39

her, passing tent pegs and poles and feeling frustratingly feeble. He was suffering with an infected toenail. The doctor at Beltana had lanced it and ordered him to lay up for three days while bathing the toe in hot salt water every two hours. It was no way to start a field trip. The ute, too, was lame. On the way up, the gear lever had slipped out of third gear every few miles until it refused to stay in at all without constant pressure. It was an ailment extending all the way back to the Mount Painter expedition—the Chev had never really been the same since.

For three days Reg relaxed in the company of his Pat. They awoke to the raucous din of white corellas and were visited by curious mobs of horses and donkeys. Even the heavens put on a display for them, with a stunning aurora australis for their own private pleasure. The only blight on their paradise was the infernal flies that harassed them from dawn till dusk.

By Saturday afternoon Reg was ready for a reconnaissance of the mine site in preparation for early morning mapping of the southern workings. On Sunday, Pat wandered the dunes while Reg traced the slabs of limestone around the old mine openings, recording the faults that gave the area its structure.

The geology of the Finders Ranges largely consists of a massive sandstone formation that has been squeezed up into great ridges and peaks. Over millions of years the limestone 'shell' covering the sandstone has been weathered away in places, leaving exposed sandstone outcrops. The Ediacara range was one such place.

It was a promising environment for fossils, Reg felt. Hadn't Mawson always believed that ancient fossils should reside in the Flinders Ranges sandstone. On their trips to the Flinders, Mawson had always exhorted his students to stay alert to the possibility of fossils. Complex animal life could not have developed as abruptly as the current fossil record suggested, he reasoned. According to the evidence so far, life evolved from single cells of blue-green algae—the oldest fossils known—to marine animals with shells, skeletons, heads and limbs, in one humongous leap. Reg shared Mawson's belief that Cambrian-aged animals with heads, brains and the means to move about, could not have appeared so suddenly. Surely there was an intermediate creature between algae and trilobites, Mawson posed. And if so, the Flinders Range was as good a place as any to look for its fossilised remains.

Reg remembered Mawson's words and let his eyes drift left and right as he scrambled awkwardly over rocks, dragging his gammy foot behind him. However, he well remembered, too, Mawson's reaction when Reg, as a 17-year-old boy, had proudly presented just such a specimen for the professor's inspection...

It was in 1936 that he had found his eurypterid fossil—the remains of a primitive 'sea scorpion' that had swam in Silurian seas 420 million years ago. In vertical sandstone beds along a dirt road up behind the Sellicks Hill Hotel, he had found delicate scrawlings and scratchings, like faint chalk marks on a sandstone blackboard. Breathless with excitement, he ran his fingertips over the almost imperceptible indents. They could be none other than the patterns made by an ancient chitinous body dragging itself over a soft, muddy surface millions of years ago, he was certain. And here was a slab carrying an impression of the actual animal that had most likely made them—a seven-centimetre fossil of a eurypterid, with a squarish head shield and a half-moon-shaped eye, a segmented body, a tufted antenna and a swimming 'arm.' He estimated that the complete animal—looking like a mixture of a trilobite and a horseshoe crab—would have been 15 to 20 centimetres long. These rocks were more than 540 million years old, he reminded himself. If his reasoning was correct, the impression of this animal that he now held in his hand, would be the oldest eurypterid fossil ever found.

Back in Adelaide he hastened to Mawson at the university, as was his habit with any geological treasure that he found.

'Rubbish,' Mawson scoffed when Reg showed him his prize. 'Just fortuitous markings.' He refused to even entertain the possibility that Sprigg could've stumbled upon a fossil of the kind that he himself had searched vainly for throughout South Australia's hills and plains.

Reg slunk away from the professor's scorn but he didn't give up faith in his 'Sellicksia,' as he had named his fossil. He would keep looking, he vowed, and he would find more.

The saga didn't end there. Three years later Reg decided to exhibit his fossil to the members of the newly formed Adelaide University Geology Club. Now he was not a schoolboy anymore but a university student who had been studying biology as well as geology for the past three years. With all his accumulated knowledge, he still believed his eurypterid fossil was genuine.

Reg began his presentation by summarising the work of another scientist who had purported to find fossils in local rocks similar to those in which Reg had found his specimen. In that case, Reg agreed with the scientific majority that they were not fossils at all but merely inorganic artefacts—random patterns the result of mud-pellet impressions, mineralized cavities or some such. However, the general idea of there being animal fossils in such ancient rocks was one that Reg subscribed to. Although it was a theory that was hard to defend without any proof.

All fossils found throughout the world so far were in Cambrian rocks or younger—the youngest layers of the earth's surface. When the Cambrian rocks were being formed 540 million years ago, so were the earliest vestiges of the multicellular life forms that would eventually give rise to the diversity of plants and animals on earth today. When older rocks—the Precambrian rocks below the Cambrian—were being formed, algae and single-celled bacteria were thought to be the only living things on earth. Scientists pondered over the great leap from bacteria to relatively much more sophisticated animals like trilobites, snails and sponges, but there was no universally accepted explanation for this tremendous gain in complexity. Some speculated that jellyfish-like creatures were the intermediaries; soft, watery 'bags' of cells with a rudimentary animal form.

If primitive worms and jellyfish had flourished in the Precambrian seas, however, the likelihood of finding evidence of their preserved remains was exceedingly small. Surely these slithery masses of cells would have disappeared forever without a trace; their signature never to be found in the fossil record. But now, Reg announced to his audience, he had found this eurypterid fossil in very, very old Cambrian, if not Precambrian rocks. It was the first evidence of advanced animal life in such ancient rocks, he declared.

Perhaps evidence of other undiscovered life forms in these rocks would now come to light, he proposed.

From the back of the room, again Mawson scoffed. This 'fossil' was nothing of the sort, he said, merely spurious markings in a piece of sandstone.

It was not Mawson's habit to attend the monthly meetings of the fledgling Geology Club. Such pursuits would distract students from their studies, he perversely proclaimed. But upon hearing of his student's proposed lecture topic that evening in 1939 he had made a point of being present—to publicly pour scorn on this ridiculous infatuation of Sprigg's.

'If he doesn't take care, Sprigg will end up as mad as David and Tillyard,' Mawson sneered to the audience as Reg left the room to prepare supper.

It was Mawson's former professor, now *Sir* TW Edgeworth David who had found the disputed fossils Reg had discussed earlier in his talk. In a similar manner to the respect Reg felt for his professor, Mawson had felt the same for his. They had climbed Mount Erebus together in 1908 during Ernest Shackleton's British Antarctic Expedition and maintained a friendship till David's death in 1934. Mawson had even named the Radium Hill radioactive mineral davidite after him. It had been scientifically impossible though, for Mawson to join with David in touting his supposed fossil finds.

In fact, the whole episode had gotten Mawson's dander up. He had been away in England raising funds for his third expedition to Antarctica when the news broke. On the 10 May 1928, David had announced in a talk to the Royal Society of South Australia that he had discovered fragmentary remains of large worms and arthropods—segmented insect-like creatures with jointed legs and an external skeleton—'of quite extraordinary interest.' The specimens came from 600 million-year-old limestone rocks at Reynella, south of Adelaide, not far from Sellicks Hill.

'There exists near Adelaide a very treasure-house for students of evolution,' David pointed out. He hoped, he said, 'that young palaeontologists in South Australia may be encouraged to work in this very difficult and yet most fascinating field.'

A month later he presented his results to the Royal Society of New South Wales, following which, newspapers in both Sydney and Adelaide applauded 'Adelaide's science treasure' and recounted what a 'titanic surprise' the discovery had been to David after 30 years of searching.

'The principal scientific interest of that important discovery is its bearing upon evolution, which in the light of the new world Sir Edgeworth has revealed, must place the birth of life millions of years earlier than science had supposed,' *The Adelaide Register* of Thursday the 7 June 1928 reported.

The news spread to England and congratulations on Lieutenant Colonel Sir Edgeworth David's spectacular discovery rolled in. Mawson, however, reading the paper in London, felt peeved. David's discovery made him look inept, he believed, for not making such a noteworthy discovery himself when it was reputedly sitting right under his very nose. He also felt that David was poaching on his own scientific turf.

'I am wondering what you are including in your operations in South Australia,' Mawson queried David in a letter written aboard ship on his way home. 'I have clearly explained to you by references on several occasions, both verbally and by letter, that for years past I have had a plan of campaign laid for the eventual elucidation of the stratigraphy of the older rocks of South Australia and it has been going forward according to plan…I have spent money in travel to see formations in South Africa and America and familiarise myself with glacial petrography and the character of the fossil formations of the plant and animal life of the oldest terrains. This problem I particularly decided upon as a major life work…

'…From reports in the English papers it was not at all clear what you had discovered,' he added in a later letter, 'though it did indicate that we in Adelaide had been sleeping on the most stupendous museum of clearly preserved fossils.'

When Mawson arrived home and was able to see the famous 'fossils' for himself, he believed David was making 'a mountain out of a molehill of evidence.' The fragments that David had discovered with the aid of a reading lamp and an old microscope were less than one centimetre long. In his reports he had drawn complicated diagrams with arrows indicating how these tiny pieces could be moved around and fitted together like a jigsaw puzzle to form the outline of a eurypterid-like animal 35 centimetres in length.

'Some of them, in fact are practically big insects of somewhat lobster-like habit,' David had written to a colleague. 'Their size astonishes me, being of the order of from one foot to 18 inches in length. The strength of their biting limbs must have been prodigious. Their swimming legs were very powerful.'

That Professor David with his 70-year-old eyes was gleaning all this from minute slivers of limestone viewed under a microscope, left many in the scientific world increasingly incredulous.

When his academic paper on the 'fossils' was ultimately rejected by the Royal Society of London in 1932 for being 'entirely without scientific value,' David was crushed. He believed in his fossils right to his death, considering it, 'the only original discovery I have ever made, of any consequence, after some 40 years of careful patient work.'

IT WAS ON the horns of this controversy that Reg now found himself. Mawson's words, '...Sprigg will end up as mad as David...' rang in his ears as he left the Geology Club tearoom that night. He felt humiliated and dispirited. All that was left, was to deposit his Sellicksia in the third year laboratory display cabinet where it was hidden in a drawer and assigned to the category of 'problematica.' Even for one as confident as Reg, two rejections from Mawson—one private and this one very public—were enough...

ALL THOSE TUMULTUOUS events of seven years ago and beyond, came to Reg's mind as he roamed the Ediacara hills early in 1946. Like Edgeworth David, he had not given up his belief in Precambrian life. Perhaps neither of them had nailed it yet, but he felt sure that the evidence was out there, somewhere, just waiting to be discovered. Had not David said, in his last presentation to the Royal Society of South Australia, 'It is [my] pious wish that no time will now be lost by the geological workers of South Australia

45

in exploiting, in the interest of world science, the priceless treasures in the Adelaide Hills and Flinders Ranges. They are hard to win, but to win the beautiful is ever hard!'

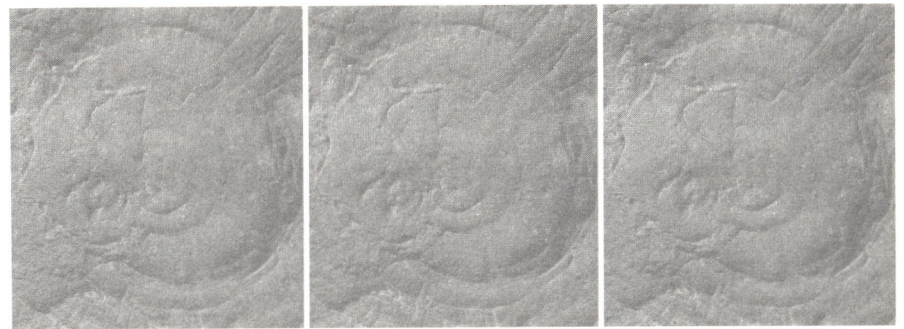

Ediacaria flindersi

CHAPTER 3

THE 40-MILE-AN-HOUR GEOLOGIST

As Reg trudged the old Ediacara mine site he kept his eyes peeled to the ground. He was at the south lode where the limestone had weathered away to reveal sandstone and quartzite underneath. The limestones looked typical of those found throughout the Flinders Ranges, he noted, meaning that they were Cambrian in age. The sandstones underneath were Precambrian then—more than 540 million years old. They were arranged in flaggy layers. In many places the burnished-orange sandstone leaves had peeled open like the pages of a book. And on one of those loose slabs he saw something that made his scalp prickle.

It was a circular imprint about 10 centimetres across. Tilting the piece of rock away from the harsh sunlight, he could see several inner concentric circles and radiating lines, and even a flattened bell-like impression in the centre, suggestive of some kind of mouth structure. It looked for all the world like the pattern of a jellyfish…but could it really be. Could he have found the mythical ancient jellyfish that filled the evolutionary gap. Did he dare hope? There were even queer markings nearby that could be the tracks of this or other creatures. He pocketed the specimen and turned his attention back to the limestones that rested above the sandstone. It was now more important than ever that he confirm their age.

One certain sign would be the presence of fossilized *Archaeocyatha*. These 'archaeos,' as they'd been nicknamed, were sponge-like animals that had flourished in spreading reefs on the sea floor in Cambrian times, when a great inland sea extended northwards into ancestral Australia. Their dainty vase-like skeletons were familiar to South Australian geologists and served as a recognised reference marker for Cambrian-age rocks.

When Reg spied the characteristic circles and curves of these 'ancient cups' coursing through the dusty rose limestone in several distinct bands he felt a new surge of exhilaration. He drew the piece of sandstone from his pocket again and gazed intently at the smooth concentric circles on the rough, sandpapery surface. He had found a fossil jellyfish. He was sure of it.

As he raised his eyes to the horizon and focused on the low, flat salt pan of Lake Torrens wavering into the sky in the distance, he realised that Ediacara was the perfect place for such a find. Here, 540 million years ago the jellyfish would have floated in a wide, warm ocean, drifting in the calm, shallow water that spread from Kangaroo Island in the south to the MacDonnell Ranges of Central Australia. When they died they would have washed-up on the tidal flats and dried out in peace, undisturbed by as-yet-uninvented predators such as fish or seagulls. Fine layers of protective silt from slow-moving estuary streams would then have slowly settled over the ancient animals, infiltrating every indent and crevice, and preserving every detail. As the burial site was covered with every successive layer of the earth's crust, the weight of many hundreds of metres of sediments slowly compressed the delicate muds and sands to hard rock, and the fossil impressions were permanently entombed...Until the process began to reverse itself.

On the stable, sun-baked Australian plains where erosion and weathering was infinitely slow and gentle, animals that had been hidden for more than 500 million years gradually began to appear. And there they lay; jellyfish drying out in the sun, many kilometres from the sea, many lifetimes later.

Reg hobbled back to camp in a cloud of thought, hoping to return after lunch to search for more specimens. However, it was now decided that he should pack up and head for the old Angepena Goldfields further east on a new mapping assignment. He left Ediacara, planning a speedy return. In fact it would be near Christmas before he could visit again.

BACK IN ADELAIDE, Reg made straight for the university. Surely this time there would be no dispute; no argument about suggestive etchings, artifacts, or any other such nonsense. Reg was absolutely convinced that he'd found a genuine fossil of a Precambrian animal. But again Mawson was sceptical. This time, though Reg would not concede. The Australian and New Zealand Association for the Advancement of Science (ANZAAS) was holding its annual meeting in Adelaide that year, with well-known paleaontologists Fred Whitehouse and Curt Teichert both rumoured to be attending. He would present his fossil at the meeting, Reg decided. Perhaps experts from outside the cloistered Adelaide academic circle might provide a more impersonal assessment.

The ANZAAS meeting that August had a packed schedule. It was the first for six years, due to proceedings being suspended because of the war, and there was an enormous backlog of data. Lecture time was at a premium.

Reg's formal presentation was on the geology of the Adelaide Series, which he had to squeeze into a 10-minute presentation instead of the 45 minutes he had been promised.

'Mawson had tried to bully the President (Bryan) into giving him the whole one and a half hours instead of just introducing the subject as was intended,' Reg confided to his diary. 'Told Bryan he was summarizing his life's work whereas Sprigg was just giving results of a couple of years research. Bryan finally held him at 40 minutes and Prof actually went to 45 minutes. This left five of us three quarters of an hour between us.'

Outside of the formal lecture schedule, however, Reg found time to show his jellyfish fossil to Dr Curt Teichert from the mines department in Melbourne and Queensland palaeontologist Professor Fred Whitehouse. Unfortunately for Reg, they shared Mawson's doubts. The conference closed with no new converts to Reg's point of view and his Ediacaran jellyfish faced a similar prospect to that of his 'Sellicksia' discovery—dispirited donation to the university museum.

A fortnight after the meeting, however, Reg received a heartening letter from Melbourne. Teichert had been ruminating on Reg's fossil and now wanted to offer some encouragement. He urged Reg to revisit the locality to search for additional specimens and also advised him to publish his discovery in a scientific journal—the only truly accepted way of presenting new results to the wider scientific community. Publication provided an opportunity for the author to reflect on his results and sharpen his arguments, and for readers to ponder the conclusions and properly question their significance.

Reg needed no further convincing. He had published his first scientific paper at age 23 and now, at 27, already had two others to his name and another underway. He now set about methodically describing the fossil he had named *Ediacaria flindersi,* after the area in the Flinders Ranges in which he had found it. For him, fieldwork was over for the year, as he recovered from pneumonia, and for once he was pleased to have some time in the office. In between finishing field reports and papers he devoted his rare snatches of spare time to his jellyfish manuscript.

When his convalescence was over, and just as summer began creeping down from the north, Dickinson sent him away for one last field trip before Christmas. It was the opportunity Reg had been waiting for. His job was to escort a group of dignitaries, including his Chief and Premier Playford, around some of the department's prime interests in the north—the talc deposits at Mount Fitton, the uranium deposits at Mount Painter and finally to the silver and lead mine at Ediacara.

THE 40-MILE-AN-HOUR GEOLOGIST

It was here that he seized his chance. Leading the group down a small gully he selected a spot for lunch just metres from where he had found his jellyfish fossil earlier in the year. While the billy boiled Reg scrambled along the hill slope and within minutes returned triumphantly with more than a dozen specimens. By the time they departed he had a haul of thirty or so. In addition to more samples of his original jellyfish he had discovered impressions of palm-sized round bodies with hundreds of radiating 'gills' like the underside of a mushroom, which he named *Dickinsonia* in honour of his boss, and another curious circular print that he named, *Beltanella*, after the nearest township to the site, Beltana.

Like Reg, Dickinson was excited by the find. The Premier, however, was less inclined. 'What earthly use are they Sprigg?' he asked cynically. 'How on earth could the State benefit from a lot of old fossils? We came here to look for copper.'

But Reg was immune to contemptuous joking. For him it was finally vindication of his beliefs, and the taste of victory was sweet. One fossil could be discounted as random patterns in a rock; thirty fossils of a variety of primitive soft-bodied creatures could not be ignored. Again, Reg took his spoils to the university. This time Mawson believed.

REG RETURNED TO his jellyfish manuscript with renewed enthusiasm and by the first week of 1947 he had a finished first draft to send for comment to Teichert—he who had been the lone voice of encouragement throughout the past year. In his paper he thoroughly described all the apparent anatomical features and attempted to relate them to living classes of jellyfish. Fine diagrams of what he thought the animals most probably looked like, from top, bottom and sideways, accompanied the text.

On the 8 May 1947, Reg entered the State Library on Adelaide's North Terrace to present the full results of his discovery to the Royal Society of South Australia, just like Professor David had done 20 years before. Reg had been committed to the society since he was old enough to attend, boasting about being the youngest ever elected when Walter Howchin had proposed him for membership in 1936, at age 17. From one of the leather chairs at the back of the room, Reg loved watching the learned professors seated in the comfortable cane armchairs at the front, discuss and pontificate and sometimes even heckle and harangue their peers.

Now it was his fourth time at the podium: Standing in Sir Edgeworth David's footsteps and finally vindicating their shared belief in advanced Precambrian animal life. Reg's only regret that night was that his former lecturer, Dr Madigan, was not seated in his customary presidential wooden throne. Two weeks into the new year, on the 14 January 1947, Madigan had died of heart failure at age 57. He had suffered a heart attack during a third-year student field trip in February 1946 and never returned to work. Those students, like Keith Johns, who had given 'CT,' as they called him, a hard time in the field, instantly regretted their boyish skylarking, and visited the popular lecturer at his home. The highlight of that visit was learning that entry to Madigan's office was via a concealed panel in the library. It seemed so Sherlock Holmes.

Reg was sorry that Madigan was not there now to share his success. He had always provided great encouragement to Reg and Reg held him in high esteem. It was Madigan to whom he had first brought for identification the rocks and minerals he'd collected as a schoolboy. Madigan too enjoyed seeing Reg 'stand-up' to Mawson. He'd told Reg how pleased he was that Reg was conducting geological investigations of the Broken Hill area and 'invading Mawson's stamping ground,' as he put it. No doubt he would've been quietly gleeful to see Reg triumph in the matter of the spurned fossils.

If Reg was expecting a similar reception to that which followed David's presentation to the Society, he would have been sorely disappointed. Far from setting the scientific world afire, Reg's seminal paper describing the discovery of the oldest animal fossils in the world, sank without a trace after it appeared in the *Transactions of the Royal Society of South Australia* in December 1947. When he sent a notice of his findings to the editor of the prestigious English scientific journal, *Nature*, his request to publish was declined. It seemed that Reg had fallen victim to the hullabaloo that had accompanied David's incredible fossil finds. Sir Edgeworth was a well-respected geologist with a lifetime of experience and *his* claims of Precambrian life had turned out to be unfounded. When another geologist from the colonies—one much younger and considerably less experienced—piped up with similar claims of spectacular ancient fossils, the British were nonplussed.

In reality, the difference between the two purported discoveries was enormous. David was studying minute shapes under a microscope and claiming that they could be painstakingly rearranged to form large worms and arthropods. Reg's discovery of a variety of primitive fauna, structurally complete and readily discernible to the naked eye was much more plausible. And yet he suffered from the David aftermath.

Reg steadfastly continued with his research and published a follow-up paper in the *Transactions* in 1949. This time he strengthened his argument by describing new Ediacaran forms and a similar jellyfish fossil found in rocks of the same age in the Kimberley region of Western Australia. Once again, nothing. The second paper, too, was met with an identical resounding silence. The scientific world was just not ready to embrace the concept that multi-cellular animal life began several million years before they thought it did. Some doubted the veracity of the fossils, others doubted the true age of the rocks in which they'd been found. Others formed no opinion at all, not lifting their heads long enough from their own endeavours to even ponder the impossibility of it all. Reg could hardly believe it.

WHILE REG WOULD loved to have fully immersed himself in his jellyfish fossils, in reality there was very little time to devote to them. Uranium, not fossils—no matter how academically worthy—was the goal for South Australia and, therefore, her government geologists. Premier Playford was an enthusiastic adherent to the idea of atomic power. He felt that an opportunity had been squandered when the Commonwealth aborted the uranium search in February 1945, and had conferred with Dickinson on the feasibility of the state taking over the Mount Painter project. Dickinson agreed. The wartime survey had been intense but brief; now was the time for more detailed investigation.

'It is incumbent on South Australia, which is one of the few parts of the British Empire where uranium exists, to do the utmost to exploit her possibilities, both in her own interest and that of the Empire,' he later wrote. 'In recent years she has undergone considerable industrial expansion which has placed a severe strain on her very unsatisfactory fuel resources. Anything therefore that promises relief to this position should receive most careful examination.'

Exploratory work at Mount Painter resumed in June 1946. The camp was re-established and geologists equipped with the newly invented portable Geiger counter—emitting a series of chattering clicks in the presence of radioactive minerals—combed the area afresh. All over the state the search was on for new deposits and there was even a public bounty on offer for new finds. Rocks and minerals of all types—including some optimistically adorned with green paint—were brought to the department in the hopes of claiming a reward.

One problem facing the South Australians, however, was an absence of information about what the newspapers were soon calling, 'the most valuable mineral in the world.' Neither Dickinson nor the Premier knew how much uranium was needed for weapons or power-generating purposes and thus they had no way of knowing if either of the South Australian deposits was commercially viable. Uranium, as a commodity, was still too new and its purposes too secretive. So when Playford heard that Professor Mark Oliphant from Britain's Atomic Energy Committee would be visiting in January 1947, to advise on the new Australian National University being planned for Canberra, he invited him to tour the mining operations at Mount Painter.

South Australian-born Oliphant was an Adelaide University graduate who had made a name for himself in the rarefied world of nuclear physics at Cambridge University in England. However, it was his now-publicly revealed key role in the Manhattan Project and the development of the atom bomb that earned him name recognition in the post-August 1945 world. The use of the weapon on the Japanese cities of Hiroshima and Nagasaki had horrified Oliphant and from then on he had shunned all research work of a military nature. But he remained committed to the potential of nuclear energy, convinced that it would provide electricity as cheaply as coal-fired power stations, perhaps more so. To him, nuclear power held the key to a new golden age of development for mankind and he was anxious that Australia take advantage of this industrial opportunity.

It was the height of summer when Oliphant flew to the northern Flinders Ranges in company with Playford and Dickinson, and other notables such as the South Australian Governor, Sir Willoughby Norrie. After two decades spent living in England, Oliphant was unaccustomed to the dry southern heat that assaulted him as he stepped from the DC3, and the memory of the visit seared a lifelong imprint in his mind. Half a century later, he still recalled hiking up a rough creek bed to one of the mining camps—a galvanized iron shed amidst a pleasant patch of trees.

'The temperature was far above the old Fahrenheit century and the heat seemed intolerable to me,' Oliphant remembered. 'The Premier insisted, however, that I climb with him over the ridge containing the torbernite. He had a geologist's hammer with which he knocked off a piece of almost every rock we passed to display the flakes of gem-like, yellow-green torbernite. When we returned I was both exhausted and dehydrated. Offered beer in the shed, I shook my head, unable to speak, and pointed to a large canvas water bag hanging on a branch of a tree. A young geologist in khaki working clothes took pity on me, found glasses in the shed and led me to the water bag where I drank more than I've ever drunk before or since. My saviour was Reg Sprigg.'

Oliphant sensed a kindred spirit in Reg and was impressed with his knowledge and candour.

'Among the many senior salesman with whom I was surrounded I had a fellow scientist—one who spoke my language yet understood well also that of the promoters of the mine. In particular, I learnt from him that the Premier's enthusiasm had to be viewed with some caution. He was able to tell me more about the deposit, its nature and its probable uranium content than any of the more senior geologists, and in words that I could understand. I liked all that I saw of that young man.'

By now, that 'young man'—Reg—was Dickinson's 'uranium man.' He had resumed his mapping of the Radium Hill deposit and also made frequent visits to Mount Painter to oversee operations there. Oliphant had agreed that the Mount Painter deposit was well worth exploiting and small-scale mining was in full swing. Shafts were being sunk, tunnels dug and holes drilled in East Painter gorge while prospectors, geologists and geophysicists fanned out to determine the extent of the deposit. The same was occurring on a smaller scale at Radium Hill, while Dickinson tried to ascertain which of the two locations held the most likelihood of success.

In June of the same year, the state government invited another expert from the UK department of Atomic Energy to inspect the state's deposits. It was Dr Charles Davidson's job to seek and assess sources of uranium throughout the Commonwealth for the potential use and ultimate benefit of the British Empire. His associate, James Cameron, flew ahead to conduct preliminary investigations before Davidson's arrival.

Reg had been dubious about the planned visit of this Cameron fellow—rumoured to be young and inexperienced—but actually found the Scottish geologist knowledgeable and good company when they inspected the Radium Hill workings together. The Geiger-Mueller radiation counter Cameron had with him was the first really portable unit Reg had seen. It seemed to make the radioactive material virtually leap out before them as they wandered the Hill—no tedious specimen collecting, sampling and returning to base to verify radioactivity.

The portable counter's usefulness compensated for the instrument's large size and weight. Cameron had had to sacrifice personal belongings when he had left London on a British Overseas Airways Corporation (BOAC) flying boat with the 25-kilogram Geiger counter in a plywood box the size of a small suitcase and just the clothes on his back. Upon arrival in Adelaide, Dickinson had at once to write to the deputy director of rationing on Cameron's behalf for special coupons to purchase two pairs of drill trousers, two khaki shirts, one waterproof oilskin overcoat, a tweed overcoat and a lightweight dressing gown.

When Chief Geologist Davidson joined his assistant on the ground, he gave a favourable opinion of Radium Hill's prospects but rejected the viability of Mount Painter. Davidson's pessimism, and his pedantic and

THE 40-MILE-AN-HOUR GEOLOGIST

Reg displays a box of diamond drill core samples taken from the Radium Hill uranium mine, July 1949.

fussy nature, may explain why patriotic Tom Playford took a bitter dislike to the man—enough to set a dingo trap by the side of his bed while he slept.

Reg crept in afterwards and removed it.

When Davidson returned to the UK he reiterated that he regarded Radium Hill with a higher priority than Mount Painter. He also took the opportunity to reassure Dickinson that operations under Reg's leadership were in competent hands: 'We both have a very high opinion of Sprigg's work,' he wrote, 'and I need hardly emphasize to you that the detailed geological report now in preparation by the mines department is likely to give you a more accurate picture of the potentialities than information you will get from any other quarter.'

By the end of 1947, exploration had begun again in earnest at Radium Hill. At Reg's direction a system of trenches was being dug—known as costeaning—to examine the face of the rock. Diamond drills were extracting deep cylindrical cores of rock that could be tested for their uranium content in the laboratory. Reg was in charge of it all, and his four- to six-week field trips were beginning to run into each other. It began to seem that he was spending more time camping in a caravan at the Hill than residing at home with his wife in Adelaide.

He earned the nickname 'the 40-mile-an-hour geologist' amongst his colleagues at the department as he dashed from one field site to another. In addition to Radium Hill, geological mapping had become Reg's prime responsibility within the department. Dickinson was keen to increase the repertoire of geological maps covering the state, and Reg, with his natural aptitude and enthusiasm for the work, was the obvious fellow to put in charge. Soon, he had several maps underway at once as he zipped from Radium Hill to the northern Flinders and down to the state's south-eastern corner. There never seemed to be enough time to spend on each project.

'Reg was dynamic,' remembers Keith Johns, who had joined the department after he graduated from university in 1947. He was a 20-year-old in his first job and in awe of the slightly older Reg, whom he already knew by reputation.

'He was very popular with the young blokes. He was close to our age and we all thought Reg was magic. He was in and out of that office; always busy; always whistling—*The Trout* seemed to be one of his favourites. He'd whistle in and whistle out. He always seemed to be a pretty free spirit and he didn't take life too seriously. He was his own man.'

Alick Whittle, too, remembers Reg as a dynamo always on the go. 'Hells bells he was a pretty wild character in those days,' he laughs. 'You've heard the story of him chopping the mudguards off the Chev?... I worked with him in the field quite a bit. We used to camp out...we used to take one of these ancient old jeeps and our sleeping-bag and you'd have a drum of water on the trailer, a drum of petrol and a few tins of bully beef. We worked like hell all day and then at night we'd go into a creek, dig a bit of a hole in the

THE 40-MILE-AN-HOUR GEOLOGIST

sand, dig our sleeping-bag in it and sleep in the creeks. That's the way we did our geology in those days—always rushing around like maniacs, flat out all the time.'

Dickinson knew that he was a pretty hard taskmaster but he felt the rush of new horizons. The country was in the grip of postwar reconstruction and development, and he was anxious to advance the mineral wealth of the state. His background in mining and economic geology made it obvious to him that informative, publicly available geological maps could serve as a lure for mining companies looking for new prospects. Since taking over leadership of the department, this Tasmanian-born director had acquired a sense of South Australian patriotism to match the fervency of the Premier's. Many a Sunday morning Dickinson spent at Playford's apple and cherry orchard in the Adelaide Hills going over mining developments and putting his case for more resources; more staff. Together they resisted any encroachments by the Commonwealth into their mining and resource domain and instilled in the band of young geologists under them a sense of freedom and purpose.

BY THE END of 1947, with Radium Hill exploratory mining underway, his first jellyfish paper under his belt, and multiple geological maps of the state in progress, Reg had yet another exciting project on the go. Wherever he went, whatever job he was working on, his mind was always busy. He never wasted any opportunity to dig, poke and investigate the ground that he walked over; always on the lookout for an interesting fossil, mineral or rock. Even unexpected plants or animals caught his eye for what it told him about the underlying soil or past environmental conditions. Every vacation or social outing was a chance to explore new surrounds. Journeys to work assignments further afield, a chance for more 'geologising' on the way there and back. He read up on the achievements of the geological 'greats'—Howchin, David, Mawson, Madigan—and filed them away in his office and his mind. He took note of the very big and the very small; able to picture the movement of the earth's crust on a grand scale and at the same time, decipher meaning from the faintest of millions-of-years-old animal

59

impressions that remained hidden to a less avid eye. He was attuned to what the earth was telling him and understood the clues of time.

Back at the beginning of the year, one of the tasks Dickinson had assigned him was to investigate the possibility of a ships' harbour at Robe, in the south-east corner of South Australia. In May, as part of his investigations he had been invited aboard the Royal Australian Navy's HMAS *Lachlan*, under the command of Lieutenant Commander Little. The ship was conducting a survey of South Australian waters, recording echo-soundings of the coastal bottom, and its presence in the South East was a good chance to learn more about the shape of the sea floor near Robe.

Peering under the waves was a new window into crustal geology for Reg and he was enthusiastic to learn all that he could. He was particularly interested when the commander mentioned that naval vessels patrolling during the Second World War had recorded deep under-sea gorges near Morobe off the north coast of New Guinea, down the mountain range about 100 kilometres from Kokoda. Reg had heard about these so-called submarine canyons in other parts of the world. They were believed to be deep furrows caused by muddy rivers rushing across the continental shelf and depositing their load of sediments into the sea during the last global Ice Age, 18 000 years ago. In typical Sprigg fashion Reg eagerly seized this opportunity to learn more.

From the original Navy survey plans he quickly ascertained that the three New Guinea examples were opposite three modern rivers that drained the Papua New Guinea Highlands. Perhaps our own Murray River had once been mighty enough to gouge out a submarine canyon of its own, he mused. It drained Australia's largest inland river basin—the Murray-Darling river system—after all.

Right away, he asked the obliging Commander Little to follow the coastline along the 100-fathom line (183 metres deep) to find out. If such a canyon existed, Reg believed that this was most likely where it would be: where the continental shelf suddenly plunged away into the abyss, and only a couple of miles from where the mouth of the ancient river would have spilled into the sea. The sea level was much lower then—100 metres below today's level, he believed—and the gently sloping sea bottom of today was a 150-kilometre-wide plain meeting the sea front just before the steep edge of the continental shelf. River gravels being swept along in the current could have eroded deep clefts in the lip of the continental shelf as they tumbled over the edge and fell into the 5000-metre depths of the Jeffrey Deep.

Commander Little steered the *Lachlan* along the 100-fathom line from western Kangaroo Island to Robe—a distance of almost 300 kilometres—with the echo sounder blipping all the way. Just south of the centre of Kangaroo Island—just where an ancient Murray River could have meandered out to sea—the echo tracing showed a trio of sharp furrows in the shelf edge. Just as Reg thought.

It was unusual that there were three valleys very close together, instead of just one, he mused, but he was in no doubt that these were the Murray River submarine canyons. The first submarine canyons to be discovered on the Australian continent.

From the single echo sounder tracing available to him, he estimated the canyons to be at least 1400 metres deep with a canyon wall gradient of 1 in 4.7—as steep as America's Grand Canyon. The power of erosion was breath taking, he marvelled. Wasn't nature fantastic.

Once again Reg wrote up his findings and sent a second paper within the year to *Nature* in London. This time the prestigious journal accepted his results, and the publication in which he described the Murray canyons and named the New Guinea canyons Gira, Eia and Waria after their accompanying rivers, was published in February 1948.

As REG'S GEOLOGICAL horizons continued to widen, his personal horizons were about to follow suit. His Chief, Dickinson, was planning a short business trip to London in 1948 and had applied for Reg to accompany him.

Before they'd even begun organising the details, however, Dickinson withdrew and decided to send Reg in his place instead. Premier Playford thought that was a grand idea. Reg could include North America in his itinerary and obtain first-hand information on uranium investigations overseas. Much was still unknown about this new wonder mineral and although it was being avidly sought all over the world, useful information regarding its extraction, treatment and application was scarce and largely still shrouded in secrecy. Governments in most countries were moving to legislate that uranium found within their borders was the property of the nation and excluded from private exploitation. It was something that everyone wanted control of though few really knew what to do with. Australia's involvement in the wartime uranium search gave it an edge over many other countries, but despite having known deposits, the path to industrialization was far from clear.

South Australia's two deposits were suspected to be of a lower grade than those in the Belgian Congo and at Great Bear Lake, Canada, the latter of which had supplied the deadly atom bomb payload, but the grade of ore that was considered to be economically viable, was still unknown. Dickinson and Playford both thought that Reg was the best person to try and snoop all this out.

For Reg, the trip would be a welcome distraction. His five-year marriage was foundering and he welcomed the opportunity to escape the foreboding atmosphere. The palaeontologist, Curt Teichert, who had expressed early enthusiasm for Reg's Ediacara fossils had, a short time later, found himself experiencing the same feelings in regards to Reg's young wife. Secret liaisons and eventual denouement followed. The overseas trip would be a perfect antidote to this bubbling cauldron of domestic discontent.

With Reg now appointed the Department of Mines delegate, geological mapping was included in the objectives of the trip and the itinerary expanded to include visits to universities and mapping institutes throughout America, Europe and the United Kingdom. Stopovers at the Emperor Gold Mine in Fiji and to the water supply board in Honolulu were gradually added to the agenda, until, somewhere between agreeing to go and actually departing, Reg's expected four-month trip to London had expanded into a nine-month international tour. He left Adelaide on the 28 March 1948, and it would be January 1949, before he again set foot on Australian shores.

Cyclomedusa

CHAPTER 4

Radium Hill

Reg disembarked the *Orcades* in Melbourne with impatient feet. Much had transpired in his nine months away and he was eager to take up the reins in Adelaide once again. His informal lecture circuit of the British Isles—in which he'd espoused the opportunities offered in South Australia in an attempt to lure experienced geologists to the dominion to ease the postwar labour shortage—had been no mere rhetoric. He believed Australia was full of potential and he was itching to get out into the field once again. He was brim-full of new techniques and knowledge to share. Nine months of networking, absorbing and lecturing was too long 'in the office' for a man most comfortable tramping about the countryside with geological hammer in hand.

Dickinson had kept him abreast of proceedings at Radium Hill while he was away and the South Australian uranium mine seemed set to move to the next stage. From his recent discussions with Mark Oliphant and members of the British Atomic Energy Commission Reg was optimistic that Radium Hill could yield uranium in the amounts and grades required to be economic. Having seen the big mine at Great Bear Lake on the Arctic Circle in Canada's Northwest Territories, and toured smaller operations in Colorado and Cornwall, he was certain that South Australia's deposit was on par. He had recorded all his observations and conversations with mine managers and geologists in his daily diaries, copies of which he had sent home to Dickinson and the Premier via secure diplomatic channels, but of course Dickinson was anxious to have a lengthy discussion in person. Reg Sprigg was now the most knowledgeable uranium geologist in Australia, and his boss had some catching up to do.

Reg took the overnight train to Adelaide and resumed his place in the Geological Survey with pleasure. But, too, there was a reason other than work, for his spirit of optimism and hope. Late in his trip, during a field excursion offered in conjunction with the International Geological Congress being held in London, he had met a captivating young lady—a 26-year-old Scottish lass named Griselda. She was spirited, feisty, glamorous and articulate, and he couldn't stop thinking about her. The well-to-do daughter of a dental surgeon, she exuded an air of poise and breeding, but her imperious expressions could also hide a sarcastic wit and a sharp sense of humour. She was slim and seductive and charming. And Reg was smitten. In his mind she was already, 'my Griselda,' and his letters to her soon followed suit.

They had met on the Island of Arran, off Scotland's west coast. On a rotten day of dripping Scottish fog, Griselda had proved a warming glow that had sustained him ever since. *He* had been tramping about fruitlessly looking for geological formations in the sodden heather. *She* was holidaying with fellow nurses in company with her mother. Their lives collided in the dining room of the Ennismore Hotel.

It was Griselda that noticed him first. Her girlfriends pointed out the sneezing, fair-haired bloke with the blue eyes and weird accent, and dared her to arrange a date. Five quid was the wager and Griselda was game. The fellow was clearly suffering with a cold, but there was a jolly air about him and Griselda was intrigued. As he took his dinner with the conference delegates she fetched her brandy flask and formulated her plan. When the group emerged from the dining room she singled Reg out and introduced herself—a vacationing nurse with free medical advice and a tot of brandy. As the rain sheeted down outside, Griselda's charms went to work and before the night was through Reg had invited her to drinks at The Douglas Hotel the next afternoon. The bet was won. But even as she introduced her trophy to her friends and collected her winnings, Griselda knew that the thrill of delight she was feeling was one of anticipation for the following day's rendezvous. The joke was already irrelevant; the die surely cast.

After departing the island, Reg had managed to visit Griselda at her home in Paisley but with his packed work programme it was impossible to

Reg took this photograph of 26-year-old Griselda when they first met in Scotland in 1948. 'Echos of Arran,' he wrote on the back

schedule more time together. He visited mines, universities and government departments in Cornwall, Zurich and London before the year was out, and on the 14 December, caught the boat train to the Tilbury docks with a minute to spare. At eight o'clock in the evening the *Orcades* set sail and Reg retired to his cabin for three weeks of enforced rest aboard ship.

It had been Dickinson's idea that Reg should sail back to Australia rather than return by air. He knew that the entire geological junket had been no kind of holiday. The South Australian Government was paying for this trip and they wanted their money's worth from their man. The first four months he had crossed the North American continent from Vancouver on the west coast, to Ontario, Minnesota and Chicago in the middle; down to Boston, New York, Washington DC and Pittsburgh in the east; and then back up to Montreal from whence he had flown to the United Kingdom at the end of July to begin a similar campaign in England, Scotland, Cornwall, Wales, Switzerland and Norway.

'Our itinerary is so full that it's all we can do to get enough sleep,' Reg had written to his Chief from abroad. 'Still we're standing up to it very well. These blighters over here are so kind that we feel quite over-entertained. They give us all the information we seek and when they're not showing us mines or plans in the evening they insist on quite a bit of steady drinking… For a time we rarely got to bed before 1.00 am and always up by seven if not six. We certainly got tired after two months of that!'

The programme meant that Reg and his travelling companion for the first part of the trip—a geologist from the CSIRO in Melbourne—were always on the move. In Minnesota, when the bus they were booked on burned down before they boarded, the pair had to hitchhike 160 kilometres followed by a 110-kilometre taxi ride to reach their next destination. On a hair-raising visit to a gold mine at Yellowknife in Canada's Northwest Territories the small Fairchild monoplane Reg was travelling in took three runs to get off the lake surface in a 100-kilometre per hour crosswind and flew the remainder of the journey at less than 300 feet to avoid low cloud. After almost crashing into a derelict mine poppet and being 'thrown about like a cockle shell in a rough sea' Reg was relieved when the plane finally made a safe landing. On the shores of Great Bear Lake, Reg was detained at the township of Port Radium while the authorities checked and rechecked his credentials before he was grudgingly allowed a brief visit to the infamous uranium mine. In this Cold War-era of secrecy and suspicion it seemed that no one could be trusted.

Despite the minor hiccups, though, the trip was an overall success and Reg was keen to take advantage of all his new experiences. Back in Adelaide, he was soon on the road again. Dickinson had appointed him head of the new department of geological mapping and the south-east of South Australia was one of the first areas under investigation. The state government was interested in the possibility of building a ships' harbour at Robe. Also Reg was enthusiastic about the oil potential of the area. When American oil expert Frank Reeves accompanied him there and expressed similar sentiments, Reg proposed a joint exploration project between the departments of mines in South Australia and Victoria, which was soon instigated.

Of course his Radium Hill responsibilities still held sway and Reg spent much of the year tripping back and forth the 460 kilometres from Adelaide to the mine site. In July 1949, with Dickinson ill, Reg played official Department of Mines consort to Premier Playford and Liberal Party leader RG Menzies when they visited Radium Hill as part of an electioneering run through outback South Australia. Reg had met both men before. Playford liked to keep close tabs on his mines department and would often summons Dickinson and/or his senior staff to Parliament House for an update, or drop by unannounced to the Flinders Street offices. He also regularly visited field operations as a kind of working holiday where he would quiz his geologists for exciting new developments to announce in his weekly 'fireside' radio chats.

Menzies, Reg had also met on several occasions, but their acquaintance had been renewed just recently when they had found themselves fellow travellers on the *Orcades'* maiden voyage from London. They would sit on deck and chat, and Menzies had told Reg how he had considered becoming a geologist before pursuing a law degree. Reg found him an excellent

Reg guides Liberal Party leader Bob Menzies (left) and South Australian Premier Tom Playford (right) through the underground tunnels of the Radium Hill uranium mine, July 1949.

raconteur with a grasp of history and the classics that made Reg feel quite culturally inferior.

The proposed whistlestop tour of South Australia suited the two politicians well. Playford assured Menzies that the South Australian community responded positively to mining and Menzies's 'endorsement' of the operations at Radium Hill, Mount Painter and the Leigh Creek coalfields would serve him well in the upcoming December 1949 federal election. In return, Playford would have plenty of time with the aspiring Prime Minister to peddle his state initiatives.

Reg rode with the Premier's official driver to greet Playford and Menzies—flying from Adelaide—at Port Lincoln, where Menzies officially inaugurated his campaign in the Liberal and Country Party stronghold. The next day, Playford commandeered the official black Buick, inviting Menzies, Reg, and the secretary of the Liberal Party, to join him, and relegating his chauffer, Peter King, to the chase car carrying Mrs Menzies, her companion Miss Wilkinson, and two press photographers.

It was a gripping journey in several ways. Over the 270 kilometres from Port Lincoln to the Whyalla shipyards, the two statesmen hotly debated matters of governance including management of the economy, primary production, communism and industrial relations. Although members of the same political party, the two had many differences of opinion and Reg was a fascinated audience to the verbal thrust and parry. Frequently, Playford became so embroiled in an argument that the vehicle he was driving became an active extension of his emotions.

When losing ground, Playford would turn in his driving seat quite alarmingly to lecture Menzies with a wagging finger. Inevitably the car would slow to a virtual halt until Peter King, honking frustratedly from behind, would draw alongside to deliver a lecture of his own on incompetent drivers and their place on the road. The two were World War One mates so King could get away with it.

When an argument went in the Premier's favour, he stepped on the gas, rounding bends at high speed, and leaving his passengers fearing for their lives. It was undoubtedly a winning streak that had him careening along a winding, ill-defined bush track on the way from Mount Painter to Radium Hill a couple of days later.

'Who's driving this charabanc, Reg?' he chortled as Reg tried to warn him of a gate hidden on a sharp bend that they were approaching at an unseemly pace. Then: 'No flaming brakes!...It's a lousy vehicle they've given us,' he continued, unperturbed, as they sailed straight through in a jumble of wire.

AFTER THEIR ONE-DAY tour of the Radium Hill mine site Playford and Menzies were completely enthused. As a result of Reg's investigations over the last three years, diamond drilling, costeaning and shaft sinking were in progress and the vibes felt good. It had been Reg's job to trace the outline of the underground uranium-bearing deposits, and gradually the edges of the ore-body had been delineated. The results revealed a system of six parallel lodes—each one an average of one and a half metres, but some up to five metres wide—spreading over a distance of almost a kilometre and to a depth of at least 120 metres. The radioactive lodes were lodged in the earth rather like whole almonds embedded in a piece of nougat.

By the time of the politicians' junket, geophysical surveys to search for undiscovered lodes in the wider vicinity were also underway and small-scale mining had commenced. Ore samples from the mine were being railed by passenger train to the School of Mines in Adelaide and the CSIRO in Melbourne, where research into the best methods for ore concentration and uranium extraction was accelerating.

The Radium Hill camp, at the time, consisted of 26 men living in dormitory cubicles and tents. When Reg was on site he stayed in a departmental caravan that he had towed up there previously and parked beyond a rise about a kilometre from the mine. There were no suitable sleeping accommodations for the Playford party, which continued 170 kilometres south to Peterborough in the late afternoon and returned to Adelaide the following day.

At Radium Hill, besides the departmental mess tent, facilities for the workers were non-existent and Menzies worried about the lack of academic stimulation for the miners. When he returned to Canberra he packed up a case of books and dispensed them to the Hill by passenger train, and thus the Radium Hill library was born.

BACK AT DEPARTMENT of Mines headquarters, the moment of truth for Radium Hill was approaching. Dickinson had to decide whether the mine was really a viable project for the public purse. The Premier was keen for Radium Hill to lead the way for Australia's entry into the new atomic world.

He wanted state management and ownership under an umbrella of British cooperation. It was Dickinson's job to advise the Premier, however, and he wanted to be sure that a mine at Radium Hill would be worthy of a seat at the world uranium table. He directed Reg to write report after report, which he then distributed to Oliphant and other members of the UK and US Atomic Energy Commissions.

'I would greatly appreciate your comments on the enclosed summary prepared by Mr Sprigg,' Dickinson wrote to the US Commission in late November 1949. 'What I would like to have from you after perusing the enclosed documents is whether you think I am on the right track as far as our venture is concerned. I know it is difficult for you to give a precise opinion but in view of your knowledge of the many sources of uranium in other parts of the world I would like to know whether the expenditure we are incurring is along the right lines.'

Responsibility for the development of a uranium mine was a huge undertaking when the market for the metal was still largely unknown. Other than Reg, there was no one in Australia with whom to consult and Dickinson relied heavily on reassurance from overseas.

On his visit to North America, Reg had learned that the Americans were pursuing every possible uranium lead. Ore deposits that contained an average of only 0.05% uranium were considered to be worth mining, with the uranium price at US$3.00/lb (less extraction costs). Ore of 0.5% uranium was labelled high grade. Early indications were that the uranium at Radium Hill contained 0.1% to 0.5% uranium and that there was about 700 tons of it in the ground. Members of the UK Atomic Energy Commission, including Professor Oliphant, felt the results encouraging enough to proceed to the next phase of development.

THROUGHOUT 1950, REG commuted between Radium Hill, Adelaide and the south-east coast, juggling his mining and mapping responsibilities.

While the majority of his recent overseas trip had been devoted to uranium, Reg had also taken every opportunity to visit geological surveys and mapping institutions in Europe and the Americas. One benefit to come from the war was the availability of aerial photographs. Working on the wartime Mount Painter project he had seen his beloved Flinders Ranges from above for the first time, courtesy of aircrews from the nearby Royal Australian Air Force base at Port Pirie. The photographs gave new perspective to the dark hulk of Mount Painter and the slanting seabed layers of the more southerly peaks. Geological observations could be recorded directly onto the photos, which were then overlapped to compile a representation of an entire area. It was a whole new—and more efficient—way of mapping.

During the Mount Painter project—when nationwide services had been at Dickinson's ready disposal—the Army Survey Directorate in Melbourne had compiled the necessary maps. There were no such facilities in South Australia. So when wartime cooperation ended, Reg had urged Dickinson to establish a mapping division of their own within the department. Reg, with his natural flair for following the earth's dips and strikes, was the logical choice for its leader. Instead of adhering to the old pastoral maps comprising a system of county and hundred plans, they would base their maps on the military map sheets. Both men vehemently believed that regional geological mapping was the foundation that underpinned mineral development of a country. Geological maps—showing such things as areas of mineral riches, possible quarry sites for sources of good building stone, or unstable regions where construction shouldn't be attempted, for example—served as a guide to miners and developers looking to exploit the earth's riches. If the government provided the basic information, they reasoned, private enterprise would capitalise and state development would follow.

Upon Reg's return to Australia, Dickinson had quizzed him on the best equipment available and then released him into the field to begin systematically mapping the entire state, beginning with the areas known to be most important: the Adelaide vicinity for general interest; the Lower South East for petroleum prospects; the River Murray Basin for coal and water; the Middleback Ranges for iron ores; the Olary region for minerals; the Leigh Creek area for coal; and the North-Eastern Flinders Ranges for minerals. The Adelaide sheet—coloured to comply with American standards and complete with contour information and creek lines—was finished first, at a scale of one inch to one mile. Mapping in the South East's Millicent, Kalangadoo, Beachport, Robe and Kingston districts consumed the first half of the year, with trips to the Flinders and outlying areas along the Barrier Highway to Broken Hill, such as Outalpa, Ballara, Olary and Kalabity, occupying the winter months and on into the spring.

At Radium Hill, while diamond drilling and the digging of exploratory shafts continued, Reg turned his attention to township amenities. Dickinson was planning a new housing development to accommodate extra workers and their families. Reg marked out garden areas where a revegetation programme could begin in an effort to combat the dust that invariably clouded the industrial and living areas. He also selected a suitable airstrip site and determined where fencing was needed to contain operations and satisfy the new security regulations. McCarthy-era anti-communism hysteria was reaching even as far as the saltbush plains of Broken Hill and restrictive security measures were making their inexorable way over the dusty horizon. The Russians had just exploded their first nuclear bomb and Eastern Bloc paranoia was running rampant.

An area of nine square miles had been excised from Oulnina station's pastoral lease and declared the Radium Hill enclosure. This area contained the uranium site and an existing dam, in addition to ample provision for mining expansion and all envisaged industrial and township facilities such as housing, mine mills, recreational facilities and the airstrip. The new fence would exclude the general public from Radium Hill itself, plus about 100 square-miles of the surrounding countryside, for security reasons. Only those with written authorization from the department would be allowed to enter, and taking photographs was forbidden. Warning notices prohibiting entry were posted and later a no-fly zone lower than 2000 feet would take effect.

Unbeknownst to him, Reg had already caught the security-mongers' eyes and by the first week of the new year, 1950, his phone calls were being tracked and he had his own personal security dossier with ASIO (Australian Security Intelligence Organisation). This divorced scientist, '…well-built with a long nose, fair complexion and sandy hair slightly thinning at the front, and who sometimes wears a bow tie…has a more up-to-date knowledge of uranium deposits throughout the world than any other geologist in Australia,' the file noted. It was this—in combination with what they believed to be his communist sympathies—that had the bureaucrats worried.

During the war, Reg had been a member of the Australian Association of Scientific Workers and in 1943/44 he had been secretary to the Association and one of South Australia's two delegates to the Federal Council. At that time the Association, which had counterparts in other countries such as England and South Africa, was concerned with the question of employment preference for returned soldiers and the transfer of scientific workers from war-time to peace-time projects upon the conclusion of hostilities. The Association hosted public discussions on a wide range of topics and meetings were generally well attended. In fact, Dickinson, himself, had addressed the Association in June 1947 on 'Recent advances in geological science.' Others in the department regularly attended the meetings, including Maude McBriar.

'It was the first attempt, like ANZAAS was, to talk about the social responsibility of science and to think in terms of community problems,' she says. But Australia's security experts saw it quite differently. To them it was a loose band of 'reds' and 'pinks'—communists and communist sympathizers, respectively. The Association's unionism and equal opportunity rhetoric smelled too much like communist propaganda and there were suspicions it was an informal communist recruitment agency. The group's activities came under increasing surveillance and state branches outside South Australia found themselves the focus of attack.

'When it was foreseen that this was going to happen in Adelaide too, the Association just folded up,' says Maude. 'It was a pity because in terms of community work and opinion, and scientists taking some role and responsibility for their work and how it affected the community, it was an important group.'

By 1949 the Association was officially disbanded. But Reg's earlier participation had already rendered him a marked man. The Director-General of Security, Mr Justice Reed, discussed his concerns with the Premier, after which Reg was summoned to Playford's office. According to the ASIO report, Playford asked Reg if he were a Communist, and on receiving, 'No,' for an answer closed the case as far as vetting was concerned.

But Reg's denial wasn't good enough for ASIO. They kept him under surveillance as a 'scientist of counter-espionage interest' for the next 10 years.

By the middle of 1950, the Radium Hill Project had been cleaved off as a separate branch of the Department of Mines and it was official policy that any employee engaged in uranium research and production had to be vetted by the state security office. Of the first list of 15 names submitted to the office for a full records check, all were immediately cleared except for Reg.

While the security issue bubbled in the background, slowly gathering steam, Reg's mind was occupied with more personal matters. On the 20 June his divorce had become final and Reg was now certain of the course he would take. It had been 18 months since he had last seen her face but the warm imprint Griselda had made on his heart had intensified on every day that had passed. Their romance had blossomed over pages of letters and thousands of miles. He longed for her now to be by his side.

When Griselda took the phone call, while on duty at Scotland's Royal Alexandra Infirmary, she couldn't say yes to Reg's proposal quickly enough. She had already found herself daydreaming about life in Australia, and paying close attention to the promotional films designed to entice emigrants to the colonies. When the diamond ring arrived in the mail shortly thereafter she knew their marriage would really happen.

It was a time of mixed emotions for Griselda. She yearned to be with Reg, and yet despaired at leaving her mother for the other side of the world. Their summer trip to the Arran Island would be especially poignant this year, knowing it to be their last. Since her father had died in 1946, Griselda had maintained the family tradition of vacationing on the island, returning with her mother to the source of so many happy memories. She had never forgotten her childhood delight at riding in a handsome cab to the Paisley railway station and then boarding the steamer to cross the Firth of Clyde. On the island they would spend their summers in Whiting Bay on the east coast and renew acquaintances with other families who returned year after year. And now, of course, the island's significance had doubled, as Griselda fondly recalled the night she had met Reg in the Ennismore Hotel.

Griselda's mother was philosophical about life, however. Her eldest daughter, Margaret, was married and living nearby. Griselda, too, deserved a chance to forge a happy life for herself.

'Dear Reg,' she wrote in July, 1950. 'I cannot say the news of Griselda's engagement to you was a surprise to me. All the airmails arriving and her delight at receiving these had prepared me for, maybe, such an event. My only regret is the distance we will be apart, however, I hope to be spared to accept your invitation and sail out to see you both settled in your home...Griselda has been a marvellous daughter to me, especially since her Dad's death, and my only concern is for her happiness in the future. I know she feels assured of it on her engagement to you. Her choice of you as her future husband has my blessing and full approval, Grace Paterson.'

Reg was elated that Griselda had agreed to become his wife. His only worry now was that she did not fully understand the true nature of his work. Most importantly, that she would underestimate the extent of his unavoidable absences from home. He believed that his continual field trips had been the reason for the demise of his first marriage and he was most anxious that there should be no repeat. Retiring from the field to an office job was not even a consideration for him, so he went to great pains to warn Griselda that he would be stationed in the outback for weeks, if not months at a time, and she would frequently find herself 'husband-less.' Griselda professed to accept that that was the way of it with geologists. But Reg worried nevertheless.

Adding to his angst was a recent job offer from Maurie Mawby, the General Manager of Zinc Corporation. The position of Chief of Regional Exploration came with a car, a house, and an annual salary of $20 000—three times his current pay. The money meant little to Reg but he knew that Griselda was used to a privileged life—maids, a cook and the like—and he fantasised about the accoutrements he could offer her on such a princely wage. Further opportunities would also flourish once he was on board with a world-class mining company like Zinc Corporation. The big 'But,' however, was that he would be posted in Broken Hill…although there was mention of a 'possible' future transfer to Melbourne. He couldn't imagine a city-girl like Griselda agreeing to set up home in a country town on the far western reaches of New South Wales, surrounded by miners and far from Adelaide's cultural pursuits. To Mawby's surprise Reg declined the offer, and prayed that he'd made the right decision.

IN SCOTLAND, GRISELDA made preparations for her departure. Her nursing colleagues were sad to see her go. She had been at the Royal Alexandra Infirmary in Paisley for the past seven years and had risen to the position of Head Radiographer and Radiotherapist. But her friends knew that it had long been Griselda's ambition to travel. She had hoped to serve overseas during the war but the UK manpower agency had sent her to Ballochmyle hospital in the heart of Robbie Burns country instead, where she had worked in the facio-maxillary unit reconstructing faces, arms and legs for airmen scarred in the Battle of Britain and merchant seaman suffering from extreme frostbite. Now her nursing days were over and a life of unknown possibilities stretched before her. In November she boarded the *Strathmore* in London, bound for South Australia.

Griselda had enjoyed many cruises on the Mediterranean and Baltic seas before the war and was accustomed to shipboard life. She had little interest in the organised deck games but always participated in the nightly dances and looked forward to each port of call. Although it was only a year since the P&O liner had been retro-fitted for resumption of passenger cruising after wartime service as a troopship, the ship was not equipped with stabilisers and many passengers succumbed to the continual rolling motion. Griselda also felt sorry for those enduring the stuffy heat below decks. She silently thanked Reg for the fresh sea air breezing through the porthole that her first-class ticket provided.

On the 12 December 1950, Griselda celebrated her twenty-ninth birthday on board and nine days later, at 11.00 pm, the *Strathmore* docked at Adelaide's Outer Harbor. The ship was more than two hours overdue but as Griselda scanned the wharf, there was Reg—waiting impatiently; waiting to embrace her.

It was too late that night to collect luggage so they returned the next morning to clear customs. With 101 passengers sorting their belongings, the customs hall was chaotic. All around them cabin trunks and suitcases were being opened for inspection, and Griselda felt dismayed at the thought of her carefully packed wedding presents and bridal trousseau being scattered and pawed over. When she handed over 24 carefully typed pages listing the contents of her trunks, the young customs officer serving her looked aghast. He hesitated a moment, then divulged that he had just been married himself. With that he chalked the unopened trunks and cases, and wished her a very happy marriage. She and Reg gratefully hastened away before he could change his mind.

Griselda had learned before she'd left home that Australia was in the grip of drought, but nothing could have prepared her for the southern summer heat and parched air that greeted her. Water restrictions were in force in Adelaide with sprinklers banned for home gardens, car washing by bucket only and permits necessary for watering sports grounds and public parks. The price of peas, beans and tomatoes had shot up due to the recent hot spell and bush fires were menacing the hills.

On Christmas day, four days after she arrived, it was 36 °C in Adelaide.

In Korea, where Australian troops were hunkered down just north of Seoul and the government was preparing to flee the city, it was cold but clear.

In Paisley, where Grace Paterson shared Christmas dinner with daughter Margaret and son-in-law Douglas, and thought often of her youngest child so far away, a dense afternoon fog rolled in and sapped the spirit from the day.

In her new homeland Griselda yearned for the Scottish snow, but would not complain of the heat. This was her life now, and she was determined to make a success of it. At least her dreams of her husband-to-be had not proved over-inflated. He was solicitous and devoted and everything she'd hoped for, and she looked forward eagerly to their wedding day in one month's time.

In the meantime, Griselda stayed with Dr and Mrs TJ Marshall at their home in Springfield. Mrs Marshall was a geography lecturer at the University of Adelaide and Reg felt that she might provide some 'motherly' support. The Marshalls had agreed to serve as proxy for Griselda's parents, since none of her family or friends would be at the wedding. Reg had also asked one of his work colleagues, Ngaire Dolling (now Teesdale-Smith), who was technical information officer in the department, to act as Griselda's maid of honour. Reg, himself, worked mainly in the city office during January but was able to squeeze in at least one trip to Radium Hill, returning three days before his wedding. That Saturday morning he was still down at the departmental depot unloading trays of diamond drill cores until it was time to don his suit. While hoisting one heavily laden tray he slipped over, falling awkwardly on his left knee.

'Busted leg unloading core,' he noted in his diary on 3 February 1951; 'Wedding in evening.'

Griselda, too, was having troubles. Since arriving in Australia she had been assaulted by mosquitoes, precipitating a spectacular allergic reaction to their bites. An injection of adrenalin had reduced her overall swelling to the point where she could slip into her wedding gown—just. It was a strapless Peter Mann creation from London with enough whalebone reinforcement to encircle a 22-inch waist and not a fraction more. A jeweller lent a larger wedding ring for the ceremony to accommodate her swollen fingers.

Griselda's bridesmaid, Ngaire, was fascinated with what she saw of Griselda's trousseau. 'It was all beautifully made, just beautiful,' she remembers. 'She had bras with lace that went round and around. I didn't see it all. But she told me she had so many of this and so many of that.'

Clothes were important to Griselda and most outfits suited her thin frame. She had designed her wedding dress herself: white corded silk, with the strapless bodice covered by a little fitted jacket with roll collar, buttoned to the waistline. She lent Ngaire a blue gown to wear on the day, since Ngaire was saving for her big trip to England in a year's time and had told Reg she would have no money to spare on a bridesmaid's frock. 'I did manage to pour myself into one of her dresses,' she remembers.

'Ngaire really looked very sweet,' Griselda wrote home to her mother approvingly.

The wedding eve was a perfect Australian summer day. Adelaide's heat wave—which had seen it endure the hottest January since the record 1938-39 summer—had finally broken, and the forecast was for fine and mild weather in the days ahead. Griselda had asked Ngaire to spend the day with her at St Helen's, the reception house at Prospect, to be on hand for all the preparations to come, and on the Friday morning she asked her to fetch some luncheon sandwiches from town—a nice little sandwich shop she knew on the corner of Hindley and King William streets. Ngaire obliged and took the tram into the city, passing Adelaide Oval on the way, where play was just about to commence for the first day of the fourth test between

Reg and Griselda were married at Scots Church, Adelaide on 3 February 1951, assisted by best man Bruce Atterton and bridesmaid Ngaire Dolling. Griselda carried a small sprig of purple heather in her bouquet. The piper was Reg's idea and a complete surprise for his wife-to-be when she arrived at the church and was piped in.

England and Australia. It was the first Ashes tour since the Don had retired in 1948.

'Of course I was a mad cricket fan,' Ngaire admits. 'I thought, I could look into the cricket for an hour and Griselda wouldn't know. So I bought the sandwiches, and then I whipped into the cricket and stayed there for an hour or so and met a few people. Then I took the sandwiches home and I didn't ever tell her that I'd been in there. Of course next day, when we're all standing in the receiving line, someone said to me, "Oh I saw you at the cricket yesterday." And Griselda just *looked* at me.'

BACK HOME, IN Paisley, Grace Paterson listened to the cricket being broadcast from Adelaide Oval and felt a comforting connection to her daughter. How lovely it would be to be with Griselda in the late afternoon sunshine on the steps of Scots Church, instead of enduring quite the coldest winter she could remember. Four inches of snow had fallen overnight and it was now clear but freezing outside.

'We are having snow, slush, rain one after the other—no change every day,' Griselda's sister, Margaret, wrote to her two days after the wedding. 'Yesterday the barometer was the lowest it has been since 1870 and at Renfrew airport the pen which turns on the barography dropped right off the paper.

'Saturday 3 February was quite a memorable day here. Glasgow had its biggest fire of the century; Scotland beat Wales 19-0 at rugby—in a game which consisted of a succession of never-to-be-forgotten incidents (so I'm told); and Mum, Auntie, Uncle Jack, Elsie, Douglas and I celebrated your wedding at the Grosvenor on Saturday night.

'Arnott-Simpson's was gutted by the fire starting about 6.00 am. There were 200 firemen at it, and one had to be carried down after three hours on the roof, unconscious. He was frozen stiff. Damage is estimated at £500 000 and many more buildings would have been set alight from sparks if it hadn't been for the snow that covered the roofs.

'Well, Gris, I thought I would drop you a little note to let you know how things are going, although, probably at the moment you will be having such a wonderful time that you really won't be caring.'

Margaret knew that as she wrote her letter, Griselda and Reg would be honeymooning at Lorne on the south-east coast, along Australia's Great Ocean Road. It was a luscious time of re-acquaintance and discovery but even then, with his new wife reclining in his arms, work was never far from Reg's mind. He dashed off a few notes and posted them to the department, receiving a scolding from Dickinson's secretary for his trouble:

'Dear Sir Reginald and Lady Muck,' she began. 'I received one small respectable-looking envelope and one large disreputable-looking envelope in the mail this morning, each containing some most awful scribble. It is about time you realised that you can only do one thing at a time properly, and if you are supposed to be having a honeymoon, why not have one, instead of messing round and scribbling reports that will take many hours to decipher. Honeymoons, well, they are meant to be honeymoons, at least that's what I have been led to believe…'

It wasn't long thereafter that Dickinson directed Reg to Melbourne—since he was halfway there already—to attend to some details of the joint investigation they were conducting with the Victorian Department of Mines. Although Griselda wasn't to know it—and actually enjoyed some solitary sightseeing in the city—it was a telling glimpse into her future. Her husband would always be on call; anything was interruptible; work came first.

Towards the end of the month they were back in Adelaide—Griselda to the small home they had rented at Hazelwood Park in the eastern suburbs, and Reg to the office. By 5 March the honeymoon was most definitely over when Reg departed for Radium Hill. It was a difficult farewell. Reg hated leaving his Griselda when she seemed so forlorn. However there was no choice but to become accustomed to these separations, for he knew there would be many. On this occasion Premier Playford; the Minister of Mines, Lyell McEwin; Professor Oliphant; and mines chief Dickinson would be inspecting progress at the Hill. All the officials seemed to know of Reg's recent betrothal and offered good-natured ribbing and advice.

'The Premier of his own accord wished us well and knew all the details,' Reg wrote to Griselda. 'He even put the Geiger over me to see how I was standing up to it.'

Once again the politicians and scientists were pleased with progress.

'The party has been a good one and augers well for the future of the mine,' Reg wrote. 'I've taken a few feet of film and will get more tomorrow when the Premier etc visit my new ore deposits and my dam site.'

While Reg was away, Griselda caught up with correspondence from home. Her old job had been won by one of her girlfriends and there was much gossip about tensions within the radiography department. Griselda could just imagine all the goings-on. Her mother wrote to say that she had sold Griselda's bicycle to another of her girlfriends for £5 and deposited it in her savings account along with three shillings and sixpence she had found in the pocket of one of her old raincoats. Also, she had ordered a wedding cake for distribution amongst friends and neighbours. She and Margaret had posted slices in little boxes on Griselda's behalf and the recipients were most pleased. She told Griselda how she had stored away the top tier to serve as a christening cake. The baker said it would last a year and Griselda's mother didn't think it at all presumptuous to expect that there would be a Sprigg grandchild within that time.

Well, he'd have to be here to take part in the making of it, Griselda harrumphed to herself, upon reading her mother's comments.

Reg was only away for a few days on that trip but his sojourn in Adelaide was equally brief. Two more field excursions in April, and by the middle of May Griselda was completely fed up with her life of domestic solitude. She hadn't forgotten Reg's pre-marriage warnings about the extent of his travelling. She didn't resent him his fieldwork. But by the same token she'd be blowed if she'd stay in Adelaide tidying a soulless house that wasn't a home. From her perspective, there was no reason why she couldn't be out there with him. It might be primitive, but at least they'd be together.

'I'm coming too,' she informed Reg, when he next returned. 'If I don't come out there I'll never see you,' she quite correctly pointed out. Reg made a feeble attempt at protest but he already knew his wife well enough to know that she was a formidable opponent once her mind was made up. He also remembered her mother's words to him: 'Reg, don't spoil that wife of yours. I know you will be inclined to do so; she gets around everyone so well. I know she did with me. But she is worth it, for all that.'

Well, this wasn't really spoiling her, Reg reasoned. But Griselda would get her own way, nevertheless.

Radium Hill it was.

Dickinsonia costata

CHAPTER 5

Mapping the State

Radium Hill in 1951 was a small but growing township of about 140 people—mostly mine workers living in two-man cubicles in the single-men's quarters. The industrial area consisted of a powerhouse, machine shop, pilot plant metallurgical concentrator and associated administrative and community buildings. There were also four houses for senior mine staff and their pioneering families, including 11 children.

Mrs Olive Marchant, wife of the machine-shop foreman, held the honour of being the first female resident.

'Good God, Mick, what have you brought me to!' she had cried to her husband when she first saw her new home. She then howled for the next 10 days while the mercury never slipped below 43 °C.

When she had moved to the Hill in March 1950 with her two young children—seven-year-old Malcolm and two-year-old Meredith—their five-roomed, prefabricated home at the edge of the industrial area was barely finished. While Olive organised the arrangement of her furniture, and the children ogled at the local wildlife, the painters completed their work.

'The emus used to come up to the door, especially if you put a mirror up. And kangaroos bounced around the place,' she reminisced to Malcolm Harrington and Kevin Kakoschke in the book *We were Radium Hill*, 'but it was the snakes that caused the most concern. One of the first I killed was while the painter was still there. A snake slithered into the kitchen as frightened as I was. I got the snake charmer and proceeded to belt it furiously. Eventually the painter yelled out "Hey Missus, you killed that snake about half an hour ago!"'

Later that year Olive's home was open for display during one of the Premier's visits to the uranium field, so that all could admire the new facilities being provided at government expense.

'Malcolm couldn't wait for the official party to arrive,' Olive remembers. 'The house was all clean and shining, everything in its place. Finally the time had come and Malcolm took up his position by the front door. The Premier and eight other politicians began filing through the house one-by-one, followed by a string of mine officials, until finally Malcolm called out, "Mum, when are they going to stop coming through? They're going to mess up the house!"'

Shortly after Playford's visit Mrs Fred Hill joined her husband at the field and Olive had some female company and the Marchant children had a new playmate in one-year-old Judy Hill. Then came the Middleton and Roberts families, with two children between them, who took up residence in the remaining two houses.

Rennie Middleton was clerk-in-charge, responsible for pay sheets and bookwork, and soon also became the official postmaster. His wife, Roslyn, thought it a great adventure to leave the comforts of city life and migrate to the outback.

'We were all quite happy to go up there as a means of getting ahead in the Government service,' she says. 'We were the only families there for…I suppose, the first year we were there. Rennie used to work fairly long hours so he wasn't home that much but we used to fill in the days pretty well. Of course no TV in those days. We'd go for picnics, and they would occasionally have a dance in Olary and we'd all pack up and go in there. The local storekeeper would take all the children at their place and put them to bed, and we'd have a night out.'

The weatherboard houses that Ros and the others occupied were very basic—three bedrooms, a kitchen and a lounge. At one end of the back verandah was a bathroom, complete with a chemical toilet and woodchip heater to heat the bath water, and at the other end a laundry.

'Our water supply was pretty erratic,' Ros remembers. 'They had a tank up on a bit of a rise which they'd pump up from the dam, so when you turned on the taps your washing would come out mud. To get a decent wash you'd have to fill up all the troughs the night before and put Epsom salts in the water so all the mud would sink to the bottom. Then you'd scoop all the water off the top to get some clean water to wash. Of course every now and again the tank would run dry and you'd have to wait for them to fill it up before you'd have any water at all.

'For a start we didn't have any power. We had to have lamps, a wood stove and a kerosene fridge. Then we had power on but they'd turn off the generator from midnight to six o'clock in the morning. I had a kerosene fridge for a start, then they finally gave us an electric one. But of course every night the power would go off, so you'd have to be pretty careful about your supplies in the fridge. We'd only get supplies in once a week. But we managed. It wasn't all that bad.'

Ros left the Hill temporarily in mid-1951 to go to Adelaide to give birth to her second child. By the time she returned with two-week-old Michael the female population at Radium Hill had increased by one.

ON THE 18 June the new Mrs Sprigg piled into the Chev buckboard with her husband and set off, once again, for an expedition into the unknown; her journey this time far shorter than her voyage from Scotland, but her destination just as foreign to her. Reg had explained that the north-east region was gripped by drought and as Griselda gazed out the window at clumps of

Griselda sent this photo of herself having lunch in her outdoor kitchen at Radium Hill, July 1951, home to her mother in Scotland. 'Billies and soup pot on fire, and to left of fire place is the famous camp oven,' she wrote on the back.

dry bluebush and the occasional stunted mulga tree, she wondered how any farmer could coax a living from these scrubby plains. This was supposedly station country and yet she saw not a single sheep. Of snakes, however, she saw more than she desired. Reg killed five that day alone.

When they finally reached Radium Hill a dust storm was raging and Griselda struggled to make out the scattered iron sheds and buildings that Reg pointed out to her. He seemed not at all perturbed by the conditions and chattered excitedly as he pulled up alongside a caravan parked alone in a stony wasteland and announced, 'Home, sweet home!'

Griselda managed to hide her disappointment as she took in her new egg-shaped abode. The van was tiny; two beds, a wardrobe and a few small cupboards. Cooking had to be performed outside over an open fire. Reg reassured her that she wouldn't have to chop her own wood with an axe—a tomahawk would be much more manageable. Provisions for ablutions were equally basic—a long-drop toilet about 30 metres from the van and a shower bucket hanging from a tree. A galvanised iron pipe lying in the sun carried water from the storage tank to the outside 'kitchen,' delivering hot or cold water depending on the temperature of the day. Drinking water had to be trucked from Olary and cost sixpence a drum.

As Reg bustled about, unloading an assortment of camp ovens, billies, pots and a meat-safe, Griselda wandered a short distance away and considered her surroundings. The wind had dropped and the sky was turning a dusty, fiery red. She listened to the cockatoos; and then the silence; and felt calmness descend. She was in the middle of nowhere with the man she adored. Life could be good.

WHILE GRISELDA ADAPTED to her new surroundings, Reg continued his work at the mine site. His most immediate assignment was to prepare for the visit to Australia of Dr William Gross, a Canadian geologist, expert in the use of the airborne scintillometer. The search for new uranium deposits in the Radium Hill vicinity had been suspended throughout 1950 awaiting the introduction of this instrument. Previous geological prospecting within several hundred square miles of the mine site had already shown fairly conclusively that no obvious surface outcroppings or shallow lodes had been overlooked by early prospectors, but a gamma ray detector mounted on a plane could cover a huge area with much greater sensitivity. Any radioactive deposit lying hidden below the ground would be detected by the scintillometer passing over it at low altitude and slow speed. Perhaps another deposit similar in size to, or greater than, Radium Hill was still awaiting discovery.

The department was paying for Gross to come to Australia and conduct the survey for them. The Avro Anson they had hired for six to eight weeks would fly at a speed of 130 kilometres per hour, 250 feet above the ground, along traverse lines spaced 500 metres apart. It was a style of prospecting new to Australia but one that had already led to a number of new uranium discoveries in Canada. Once again, the South Australian department was at the geological forefront.

Reg and geophysicist Ken Seedsman took turns accompanying Gross in the plane to help with the radioactive monitoring and camera recordings. The survey took 27 days, including 120 hours of actual flying time, and covered an area of greater than 5000 square kilometres. It was on run thirteen, flying over a low rise two miles north-north-east of Crocker's Hill that the detector's reading went off scale, measuring a radioactive count more than six times above background levels. The plane's wheels had barely touched the ground before Reg and Ken were scrambling to depart for a ground reconnaissance. Griselda came too. In a flurry of excitement they threw the billy, a packet of biscuits and two blankets into the Chev and took off for Plumbago station about 120 kilometres to the north-west.

Nobody told her they would be camping out that night.

The dirt road from Radium Hill to Olary was well used and following that there was a graded mail track from Olary township out to Plumbago. The station owners used horses not vehicles around the property, however, and beyond the homestead the track deteriorated to a bridle trail. Continuing on to Crockers Well the Chev wound its way through station yards before rugged granite hills began to close in and the going became rocky and rough. The night was also closing in, and as the temperature fell—and kept falling—Griselda cursed their lack of camping provisions and vowed she would never be caught so ill prepared again. When they finally stopped Reg made a campfire and Ken burrowed out a groove in the sand, which he lined with one of the blankets to form a rudimentary bed. The second blanket draped over the three of them provided scant protection from the cold August night. Morning couldn't come soon enough.

At first light Ken and Reg began scouting with the hand-held Geiger counter, which soon confirmed the area of radioactivity located by the plane. It took a little longer to identify the yellowish-brown waxy mineral with a bright internal lustre that was the actual radioactive element. The two men spent the rest of the day plotting a grid over the area and pondering exactly what this new mineral could be. It was quite different from the Radium Hill davidite and Reg suspected it to be something completely new. If so, he nominated the name absite for it, in recognition of the ABS (air-borne scintillometer) that had been so instrumental in its discovery.

By October, an official work camp had been established at Crockers Well, under Reg's supervision, and further exploratory work was planned for 1952. Premier Playford was excited by the new prospect and hoped that uranium from Crockers Well would find a similar market to the Radium Hill output. When he went with Reg in December to view the site, he 'went crazy' at the absite exposure. At his urging the usual practice of halting fieldwork over the hot summer months was cast aside and heavy earthmoving work was scheduled for the second week of the new year.

As the Crockers Well development accelerated, Reg was becoming increasingly integral to the state's program of uranium mining. However, the security service had not forgotten Reg, and their persistent concerns were proving an increasing headache for Dickinson.

'With reference to our recent discussions on Reginald Claude Sprigg, further study of this file has convinced me of the dangers inherent in the situation which it reveals,' ASIO headquarters in Melbourne wrote to the regional director of South Australia in December, 1951. 'I would appreciate an early statement of the action which you then propose, with such details as you are able to supply at this stage.'

While the security chiefs admitted in their own files that there was, 'no definite evidence that Sprigg was, or is, a Communist, or that he was, or is, engaged in any espionage activities,' they felt him a threat nevertheless. They were anxious to have this 'suspected Communist' removed from any contact with uranium and/or confidential information, and indirectly pressed Dickinson for a response.

But for Dickinson it wasn't so simple. Reg was still his most experienced uranium geologist and he needed his knowledge and expertise in the field. Over a year ago he had advised that Reg had been placed in charge of the mapping division and hinted that his work on uranium research and development would diminish. But, not surprisingly, ASIO had determined that Reg was still intimately involved.

'We have to advise that Sprigg still has access to information relating to the most secret aspects of the Uranium Project,' the Adelaide branch office informed Melbourne headquarters in July 1951. 'Sprigg is consulted on problems relating to geological aspects of the work and would therefore have complete knowledge of progress being made. Further developmental work of the Project hinges to a large extent on Sprigg's professional advice, and he must necessarily be supplied with ore assay results…the most secret aspect of the Project.'

In an effort to resolve the situation, Dickinson began to restrict circulation of confidential files and to deny Reg access to certain uranium-related reports. Naturally, Reg became frustrated at this incomprehensible withholding of information and blamed Dickinson for what he considered to be such absurd ostracisation. He knew that Radium Hill information was being shared with others in the department. Why was he being victimized, he asked himself. He was the one who had resurrected the derelict adits and shafts back in 1944—crawling along on his belly and clambering down rickety ladders on his own. He had devoted much of the past five years to the mine, and now his motives were being questioned.

Meanwhile, Dickinson groomed a successor for Reg in Lee Parkin, who had returned to work at the Department of Mines in July 1950. Parkin was assigned firstly to Mount Painter and Radium Hill and then sent to America for three months to study uranium exploration in Colorado, New Mexico and Utah.

By the end of 1951, Reg felt the situation had become intolerable. How on earth could he do his job without access to his own data? When Griselda announced she was pregnant it seemed like an omen. It was time to break camp and leave the squabbles of Radium Hill behind. His wife needed a proper nest in the city and he could concentrate on his mapping again. Straight after Christmas he handed over his uranium responsibilities to Parkin and he and Griselda went house hunting in Adelaide.

EARLY IN THE new year Reg and Griselda purchased their first home—a three-bedroom house at Somerton Park, just off Diagonal Road. Vineyards at the eastern end of their street gave the area a rural feel and to the west, just one kilometre away, was the beach. It was a triple-fronted brick home with a simple front yard behind a low wooden fence separating them from the dirt street. The rear of the house faced onto a small cul-de-sac and Griselda soon made life-long friends of her neighbours in the quiet turn-about. Inside, the house was completely bare, but Griselda luxuriated in having her own—indoor—bathroom and laundry, with gleaming taps from which clean water gushed in glorious abundance. What a change from the muddy water stored in buckets she had grown used to at Radium Hill.

Since they had purchased the house outright their bank accounts were now dry, but in the tradition of many young newlyweds, they made do. Cupboards and chairs were fashioned from scavenged fracture boxes—the sturdy wooden boxes in which explosives were sent to the Radium Hill mine. Second-hand Indian rugs soon covered the pine floorboards. Griselda also began accumulating nursery furniture in expectation of the baby's arrival in June.

Reg returned to his mapping. His uranium days were not completely over—there was still the occasional visit to Radium Hill, and a trip to Darwin to view the uranium deposit that had been discovered at Rum Jungle in 1949—but mostly he could focus his attention on the geological mapping of the state.

Since his return from overseas three years ago, he had been keen to see his state better covered. Australia, and South Australia in particular with its miles of outback areas, was severely under-mapped compared to the rest of the world. One inch to one mile was the standard scale and the whole of Britain had been mapped to this level and many areas in much more detail. Only half of North America and a quarter of Canada was completed to this stage, but the figures for his own home turf were woefully far behind. The state of Victoria was the best, with 18% coverage, followed by Western Australia. Poor old South Australia languished at an embarrassing one third of one percent.

The mapping division had blossomed under his leadership with at least 11 military map areas complete or nearly complete and 21 others underway. But there was still so much to do; so many unexplored pockets.

With Radium Hill off his shoulders he could now devote all his energies to the exploration of his state. He longed, in particular, to probe the northern desert regions. Who knew what treasures could be buried under those arid, sun-baked sands—uranium, gold, iron ore, maybe even oil, although the renowned American petroleum expert, Frank Reeves, had just published a paper in the *Bulletin of the American Association of Petroleum Geologists*, condemning the possibility of oil in Australia.

'The failure to find oil in an area as large as the United States can not be attributed to a lack of enterprise or technical ability…it evidently is due to the absence of certain geological factors necessary to the accumulation of oil,' he stated, after four years of field work and a complete review of the literature. 'Australia has no commercial oil fields and because of unfavourable geological conditions will never produce enough oil to supply its domestic needs.'

But Reg was not so sure. He knew the accepted mantra—that Australia was too old for oil, its blanket of sediments too thin. There were no indicator sites where oil seeped to the surface crying out for a well to be put down, and to date, none of the hundreds of water bores that had been drilled along stock routes crossing the inland country, had given any hint that oil reservoirs like those that fed the gushers in America, lay close by. But still, there were clues.

In 1900, drillers digging a water bore on Hospital Hill in Roma, Queensland, struck gas that powered the town's lights for ten days before blocking in the well stopped the flow. At Wilkatana station, north of Port Augusta in South Australia, minute blebs of oil had been found in a water bore in 1934: Prior government geologist, LK Ward, paradoxically pronouncing it the 'least intangible' oil showing in the state. And too, there were reports of oil washing up on the beach along the south-east coast. Debate raged about the origins of the oil, or whether it was even mineral oil, but Reg had seen it with his own eyes and believed it to be genuine. At the very least it deserved a serious investigation.

It was true that some of the big international oil and mining companies like Zinc Corporation, Standard Vacuum Oil Company and Shell had conducted a cursory search for oil around the country, so far without success. But still Reg yearned to have a good look at the desert regions where his boyhood heroes had roamed—explorer Charles Sturt and surveyor Augustus Poeppel and lastly Madigan, who had crossed the Simpson in

MAPPING THE STATE

1939. At least the geology should be properly mapped, he reasoned, and maybe a discovery to surprise the sceptics would be made.

However, the north of South Australia was not a current mines department priority, according to Dickinson, and Reg was sent in the opposite direction instead, to Kangaroo Island off the tip of South Australia's Fleurieu Peninsula in the south-east corner of the state. He would be away for several days working on the geology of the Kingscote military map sheet covering the whole of the island. Griselda decided to accompany him to escape Adelaide's February heat…and learnt her first geological term in spectacular fashion.

Most of the island was composed of laterite soils that gave the unsealed roads a sparkly glint but a slippery finish. When their Landrover swerved violently into the scrub as Reg tried to overtake a tractor, he blamed it on the tyres' lack of grip on the laterite road. Griselda gave thanks that she and her unborn baby were still in one piece, and never forgot what laterite looked like.

At D'Estrees Bay on the south-east coast of the island Griselda's Radium Hill tomahawk skills were called on as they negotiated an overgrown track to an abandoned whaling station. It was her job to walk ahead of the vehicle and cut a trail through the scrub while Reg followed in the car. They finally arrived at their campsite in darkness on a cold, blustery night.

A few days later they were camped on the beach at Emu Bay at the most north-westerly tip of the island. The geology here was quite different to the cystalline granite and metamorphic rock that formed the foundation of much of the island. The sandstone beds, limestone boulders, and grey and purple shales of Point Marsden, White Point and Emu Bay had been one of Madigan's special areas of study. In some of the more slatey layers he had found raindrop impressions and obscure animal tracks that he presumed to be those of trilobites; but never a cast of the animal itself. It had been during an expedition to the area with a group of Adelaide University geology students that he had suffered the heart attack that eventually led to his death.

95

Reg now took the opportunity to search for the fossils that had eluded his lecturer and others before him. Walking onto the rocky beach he selected the most likely shale band and hit it with a sledgehammer…Then gasped in amazement as the rock fell apart and bright-orange fossil trilobites showered his feet. In a frenzy of excitement he loaded up the Landrover with them and hastened back to Adelaide to share his find with palaeontologist Dr Martin Glaessner, a new addition to the university's geology department. Glaessner confirmed that Reg had found two species of trilobites (*Redlichia* and another later to be named *Estaingia*) that had never before been found together. They were more than 500 million years old.

Reg's reconnaissance survey of Kangaroo Island brought home to him yet again how much he enjoyed being out in the field. He loved observing differences in the countryside. The rocks and soil, even the trees and vegetation provided clues as to how the earth had developed; how the surface of our planet had twisted and buckled, slumped and dipped as different formations had pushed and squeezed against each other. He loved figuring out how the layers of sediments had been deposited in response to global or local climate change, and visualising how the ordering of the layers stayed consistent as they popped up in one area, only to disappear underground and re-emerge somewhere else. He loved the way that a thin band of fossils mutely spoke volumes about the environmental history of an area—whether it had been dry or water-covered, warm or cold, calm or turbulent.

Since he was five years old he had been beholden to geology. For the past 14 years—first as a university student and then as a government geologist, he had had the privilege of official endorsement to roam the state and acquire an intimate knowledge of his home turf. He now had an innate feeling for how this giant sedimentary basin with layers up to 20 kilometres thick and that extended over 90 000 square kilometres from the tip of the Flinders Ranges in the north to Kangaroo Island in the south, from Lake Torrens in the west to the NSW border in the east—the Adelaide Geosyncline—had formed. It was time to tie it all together and lay out his treatise for how it had all happened.

He approached Dickinson about writing a dissertation on the Adelaide Geosyncline; an overview of previous research combined with a detailed analysis of his own findings and the unique conclusions he had drawn. It could be published for the mines department as Bulletin number 30 in the series of Geological Bulletins published since 1912. In consultation with his boss he also decided to approach the university with the idea that he enrol for the Doctor of Science (DSc) and submit his thesis to fulfill the requirements of the degree. It was the highest possible academic achievement and one that only four others at Adelaide University had completed before him (D Mawson, LK Ward, RL Jack and AR Alderman). At a meeting of the university science faculty on the 10 June 1952, Reg's candidature for the DSc degree was accepted. Mawson, Glaessner and Professor Eric Rudd were appointed as examiners, in addition to one external examiner, as was the usual practice.

Reg was now absorbed in fieldwork for the Outalpa map sheet covering the area immediately north-west of Radium Hill. Although he was mostly free of Radium Hill responsibilities there were still reports to complete and plans that required his involvement. The more he had to do with it, the gladder he was to be largely free of it. The security regulations had ballooned to ridiculous proportions. Playford, himself, had confided to Reg that he thought all this security business was over the top, but the Americans demanded it and he was forced to comply.

In August 1951 Playford and Dickinson had travelled to America to secure a deal with the US Atomic Energy Commission. The Americans wanted a long-term, secure supply of uranium and they were impressed with the Radium Hill proposal. They agreed to assist in further exploration work in South Australia in addition to funding developmental work at the mine site and a uranium treatment plant at Port Pirie. In March the American experts had toured Radium Hill and in April the agreement was signed. The Americans would pay most of the development and operation costs of the mine in addition to payment for the uranium produced over the next 10 years. Playford had won himself a good deal, but the Americans insisted on certain conditions in return, and strict security controls was one of them. A department security chief was appointed as a result.

Dennis O'Leary was seconded to the Department of Mines from the Melbourne office of the Commonwealth Security Service for 'special duties in connection with the Radium Hill project.' Changes were evident the minute he started. All staff had to sign an 'Oath of Faithful Service and Secrecy,' like a confidentiality agreement, and any visitors to Radium Hill had to be screened just as the staff were.

Lee Parkin ran foul of the new rules when he attempted to arrange for his wife to accompany him on a visit there. He expected her security clearance to be cleared without delay and was surprised when he was summoned to O'Leary's office and informed of a 'small problem.' It seems that one of her referees had had the audacity to attend a peace rally in Sydney, and this misdemeanour had cast suspicion on his wife's application. No matter that said referee was the minister of the local church.

Privately, many of the staff considered all this security hoo-hah one big joke. They nicknamed O'Leary 'Dick Tracy' and rolled their eyes at how seriously he took his job. They sniggered and shook their heads when blunders occurred that highlighted the farcical situation. Rennie Middleton remembers requesting a file from O'Leary in relation to having furniture shifted from one house to another at Radium Hill. When he received the file he saw right away it was the wrong one. Instead of housing development costings he could read the summary of uranium ore reserves at Radium Hill! 'I went over and looked at it and said this is not the thing,' he says. 'O'Leary was quite upset because of the security they had on that particular file—it was marked Top Secret.'

On another occasion, when Reg stopped at a nearby pub on the way to Radium Hill he was dumbfounded to be 'enlightened' on the status of the mine by a truck driver drinking at the bar. When the bloke confidently recited how many thousand tons of ore they had, at so many pound of uranium per ton, Reg knew the figures to be true. They were the latest estimates that he and the mine manager had just compiled. Reg was shocked at the accuracy of the man's information—information that was supposedly restricted to the Chief, the Minister and the Premier.

Those like Reg and Parkin who really knew the capacity of the uranium mine believed it extremely unlikely to be an espionage target. The uranium content was too low in the grand scheme of things to be worth the trouble. Still, the security department grew and became the source of increasing discontent.

To make room for the expanding numbers of staff and files, the geological survey and drafting divisions of the Department of Mines were moved out of the Flinders Street city offices to what was known as the Exhibition building a few streets away. Reg felt that geologists were rapidly becoming second-rate citizens within the organization. Radium Hill and its associated security dominated everything. Uranium mining and the new Leigh Creek coalfield were the glamour gals of the department while those that devoted themselves to the grassroots occupations of mapping, surveying and drafting were pushed into the background. He didn't like the feel of this bloated new mines department with its 700 or so people on the payroll. When he'd joined the staff, back in 1945 there had been only 81 employees in the whole organisation and the aura among the 24 salaried staff had been one of geological explorers discovering the frontiers of their country. This unwieldy conglomeration that had taken its place had an air of bureaucracy that grated on him.

His relations with his chief had changed too. Dickinson had become more distant and erratic and Reg found it difficult to engage with him as one geologist to another. Admittedly Dickinson was swamped with work most of the time but Reg began to detect an almost hostile attitude towards him. The wily Playford only inflamed the situation by goading Dickinson in Reg's presence, saying, 'if you don't do what you're told there's always Reg here to take your place.' Reg was mortified, especially as he'd never aspired to the position of chief, and had declined the position of deputy when it had been offered to him.

However Reg was not one to dawdle in the doldrums. He was a 'glass half full' kind of chap and there was always something to be thankful for. On the

22 June, Griselda had given birth to a baby girl they named Margaret. His new daughter was a delight. But as usual there was little time at home for him to enjoy either of his girls.

In August he travelled to Sydney for the annual ANZAAS meeting, and was instrumental in the formation of the Geological Society of Australia (GSA). When he returned home he chaired the final meeting of the Adelaide University Geology Club, which was then dissolved and immediately replaced by the South Australian division of the new GSA.

Christmas that year was unusually quiet. Griselda had taken baby Margaret to Scotland to meet her maternal ancestors. It was two years since Griselda had seen her family and friends, and she looked forward to seeing her mother and sister and meeting her nephew for the first time.

Reg spent the summer months planning for an ambitious new expedition. In May 1953 he would be leading a survey party into the Musgrave, Everard and Tomkinson Ranges, far out in the north-west of the state where South Australia meets the Northern Territory and Western Australia. It would be the first time in many years that geological parties had ventured into such truly remote areas. The search for uranium now ruled everything and no stone could be left unturned.

Reg accepted this new assignment with enthusiasm. Here was a chance to return to the fieldwork that he loved—being out in the unexplored spaces; soaking up all that the earth had to tell him; away from the politics and oppressive atmosphere of the department. This was the preserve of the Pitjantjatjara Aborigines, a part of the country that only a handful of white men had ever seen. He would be on the scent of explorers like Herbert Basedow and Ernest Giles, and the infamous Lasseter who had reported a reef of gold, never to be seen again.

The expedition required careful preparation. They would be operating in inhospitable terrain hundreds of kilometres from established food and water supplies or medical assistance. Running camps would be the order of the day, where three or four men with swags and the bare necessities would be always on the move. There was a lot of ground to cover.

MAPPING THE STATE

Two Landrovers fitted with extra fuel and water tanks, and a three-ton, four-wheel-drive Blitz truck were packed up and dispatched on the train north. Supplies included tents, sleeping bags and stretchers; ore bags; emergency first aid kits; 44-gallon drums of petrol; a 12-gauge shotgun and a 22-calibre repeater rifle; a dozen fly nets; supplies and spare parts for the vehicles; and four packets of boot tacks as well as a sheet of greenhide leather, a shoemaker's knife and a boot last. Whatever else they may run short of, there would be no excuse for poor footwear.

By the 22 May 1953, the seven men of the north-west survey had assembled in Oodnadatta, 1000 kilometres from Adelaide, ready for duty. The next day they would set out for Granite Downs station, a further 200 kilometres to the north-west. Reg was in charge of geologists Bruce Wilson and Ron Coates, geophysicist Bill Knapman, a mechanic, a cook and prospector Jim 'Hoppy' Johnson, who had officially joined the department earlier in the year. Somewhere en route Reg's friend, Allan Vial, in charge of aerial mapping for the Department of Lands, would be joining them with his mapping team in the Avro Anson.

Reg and Allan shared a passion for accurate mapping and were alike in other ways too. 'We got on very well together, Reg and I,' Allan says. 'We were both enthusiasts for whatever we did—get on and do it, and do it to the best of your ability. Reg would thoroughly enjoy what he was doing. He didn't look for the limelight, he put all his efforts into doing the job. And he was a good leader. He led from the front and that's the only way to go.'

Allan had been a cadet surveyor before joining up in 1942 as a navigator/observer. When he returned from the war he spearheaded the new photogrammetric section. Features such as roads and rivers on aerial photographs could be traced over with a cursor connected to a plotting machine to create the basic structure of a map. It was entirely new technology back in the late 1940s. Reg had seen the A6 stereo plotting machine in use and toured the Wilde operations in Switzerland on his overseas trip. He realised that these instruments represented the most efficient way to map and he became an unofficial ambassador for their adoption by the Department of Lands, and, he hoped, by his own Department of Mines.

ROCK STAR

Unlike some of the surveyors in his department—who feared that the associated reduction in fieldwork would put them out of work—Allan was keen to embrace the new method. His war experience as a navigator and bomb aimer, combined with his obvious enthusiasm for the task, made him the logical choice to take charge of the mapping programme. On the north-west survey, when he wasn't up in the Anson, he joined Reg's team on the ground, using his sextant and astro-tables to guide them through the desert from waterhole to waterhole by way of the stars.

Allan and Bruce Wilson rode together while Reg and Hoppy Johnson drove the other Landrover.

'Whenever we camped,' Allan recalls, 'after miles and miles of bounding through spinifex we'd stop for tea and Reg and Hoppy would go out, maybe half a mile, and start lighting fires. And every night the same thing. We called it spinifex madness. Because we'd get so tired of spinifex. The top

Continually bumping over the giant spinifex mounds around Mount Davies inspired 'spinifex madness' in the men of the north-west nickel survey, September/October 1954.

is nice and wavy, but it clumps. And in a vehicle you're just bump, bump, bump, all the time. Then every now and then you'd get an area of scrubby timber—no spinifex—and we'd hurtle through. No track or anything, and we'd be racing along. Absolutely crazy! Spinifex madness.'

Allan remembers Hoppy—named for his up and down gait due to one leg being significantly shorter than the other from polio—as a funny man. He was self-taught but had a feel for geology that meant he was soon recognised as a local authority on minerals and their occurrences. When Reg had been organizing the expedition he had specifically requested the services of a prospector, advising Dickinson, 'The person selected should be young and physically fit. Above all he must be able to mix well and cooperate with fellow members of a party which will live in relatively isolated areas for considerable intervals.'

Reg had no problems with Hoppy's prospecting skills—they were first-rate; it was his personal behaviour that caused a headache for Reg. At dinner one night Hoppy drew a hunting knife from his belt and threw it with enormous force at one of his dinner companions who had riled him. The knife lodged in the edge of the wooden camp table, millimetres from Bruce Wilson's lap. Reg threatened to banish Hoppy back to Adelaide and would have done so immediately if he had known then that the same pair had been involved in an altercation a couple of days previously. They had been out in the field together when Hoppy had spotted a bush turkey. An acknowledged expert shot, he was disgusted with himself when he missed. Twice. When Bruce served up some good-natured ribbing from the vehicle, Hoppy turned and shot the tyres out, one by one, while Bruce sat quaking behind the steering wheel.

'He was a volatile, grouchy character,' recalls Bob Laws, who later worked with Hoppy, 'but I liked him. He would fly off the handle at the drop of a hat. On one field trip he woke everyone up in the middle of the night because the crickets were annoying him, so he went outside and shot into the tree at them. When all the excitement had died down and everyone returned to their tents there was a tremendous BOOM!' Bob laughs. 'Jimmy had set explosives at the base of the tree to blow it up.'

'He wouldn't wear any socks or underwear, and in the field he would never change his clothes the whole time he was away. He'd wear very little. In the Chatham Islands—where it's very wet and cold—he agreed to wear a waterproof, but then the flapping of the plastic jacket annoyed him and he threw it off a cliff!'

WHILE REG ATTEMPTED to preserve good camp relations while keeping the survey progressing and ensuring food and water supplies were ferried out to them from Oodnadatta or Finke, he was also dealing with a constant flow of memos from his chief regarding the high food costs incurred by the party, in addition to evolving restrictions on where they could travel as Australia prepared for its first nuclear test on the mainland, somewhere in their general vicinity.

'Your reference to meal costs at £6 per person per week has caused considerable alarm here,' he wrote in reply to Dickinson after three months in the field. 'Mostly our midday meal consists of toast and butter and tea, and frequently other meals…are only a little more elaborate. We therefore find it impossible to believe that our meals are so costly…Mr Ford, the cook, is the epitomy of economy in cooking and altogether the camp personnel now detest the sight of tinned foods. I request permission to go through all accounts and charges to the mess when I return to Adelaide as I feel there is something grossly wrong.'

Even from the fartherest corner of the state the frustrations between the two appeared to be growing not diminishing, and for the first time Reg began to question his future with the department. In considering other alternatives he turned for advice to his mentor, Mawson.

'Dear Professor,' he wrote from the Aboriginal mission at Ernabella in early October, 'I do wish to thank you for your generous offer of support for my proposed application for the New Zealand Chair of Geology. You must forgive me for my mercurial indecision on this score. The mines department in SA has many remarkable advantages these days, and one is likely to forget them when being messed about by one such as Dickinson. With my long association with this man, I should know him by now, but all I find I know is that he is the most unpredictable and irrational person I know. It is sometimes exceedingly difficult to keep one's sense of humour when working with him, and even more difficult to prevent one's own exaggerated sense of instability from leading to a rash decision. For this reason I am deeply grateful to you for your wise council on Friday last when you suggested I should remain with the department to complete the North West survey…'

The winter field season was drawing to a close and the survey would be suspended at the end of the month. To date their search for uranium over 40 000 square kilometres of country had proved unrewarding, and fuel and water supplies were running low. However, there was one final foraging safari Reg was determined to make. Mawson had alerted him to the presence of 'siliceous breccias' reported by the explorer Herbert Basedow in 1905. In his reports to the Royal Society, Basedow had described rocky 'reefs' stained with green, red and yellow minerals. Reg thought the formations could be Lasseter's claimed 'reef of gold' and he was keen to investigate for himself. The general geology of the area made him quite certain there was no hidden fortune to be found, but perhaps the colourful staining flagged a uranium deposit.

Their route to 1053-metre-high Mount Davies, only 20 kilometres from the Western Australia border and the same from the Northern Territory border, took them through carpets of wildflowers following a good wet season. They marvelled at the scenery but cursed the worst giant spinifex imaginable and despaired at the paucity of minerals to be found. At last the ruined ramparts of the Tomkinson Ranges reared up before them and they recognised the massive angular blocks of the Mount Davies breccia. The formation stretched to the west as far as they could see.

There was no time for a detailed examination but enough to determine that the green stains were most likely caused by nickel deposits. It was the most hopeful discovery they had made on the entire survey. Seams of white quartz and red and yellow iron colourations made Reg more than ever convinced that this was the scene of Lasseter's fictitious find. They hurriedly collected samples and returned to their campsite at Lake Wilson to prepare for the long trek home.

The next morning, when it was discovered they had left some bags at the site, Reg returned to retrieve them while the others broke camp. As he stood alone, in the midst of ancient, untamed wilderness he heard three deep, heavy booms reverberate around the sky—the sound of an atom bomb detonating at Emu field, 450 kilometres to the south-east: The moment Australia's nuclear innocence evaporated inside a mushroom cloud.

REG RETURNED HOME after five months in the field on 29 October 1953. He returned to the news that his thesis had been rejected by the university. The examiners' reports were unfavourable and he was advised to thoroughly revise and rewrite it before resubmission. He knew that it had been typed in a hurry—Griselda banging away in her one-fingered manner for 200 pages—and that there were definitely editorial errors here and there. But he was more distressed by rumours circulating that the material was not all his own work; that there was simply too much research for one man to have accomplished on his own.

Reg felt sure that machinations by Dickinson were to blame. This really was too much. The thesis represented years of teenage tramping and solitary weekend exploring from long before he was an assistant government geologist. Yes, his work with the Department of Mines had all been folded in to his magnum opus, but still he had been largely working in isolation, and any work by those who had trod the ground before him, like Mawson, had been properly referenced. Certainly the ideas were all his and no one else's. His theory that grand movements and deformation of the earth's crust had formed the great basin structure known as the Adelaide Geosyncline, was entirely unique. Maybe its very uniqueness had been enough to ruffle feathers in the Adelaide geological community or perhaps the simmering rivalry between the university geology school and the government geological survey had been enough to undermine his efforts.

'It's true, he was the very first geologist to invoke tectonics in geosynclinal evolution,' says Dr Wolfgang Priess, who later studied the same subject in the 1980s. 'He *was* the first person to think of the geological evolution of the region. The ideas that he expressed were things that no one else had thought of before and they were based on very sound observations. It was way ahead of its time, and I've no doubt that he did it all, because he had no collaborators and there was nobody else he could've pinched the ideas from.

'It was a colossal amount of work and absolutely worthy of a doctorate *but* he couldn't submit it in the condition it was in. I could see why it was rejected; it looked like a first draft—full of editorial scratchings. It probably needed to be retyped. I was always under the impression that he overreacted to a request to tidy and polish his thesis.'

But Reg was convinced that there was a conspiracy against him. Not only had the university declined to award the degree but now Dickinson was revoking his decision to have the material published as an official departmental bulletin. It seemed to him that nowadays whatever he proposed to his boss was met with a brick wall. His investigation of the Adelaide Geosyncline seemed destined to languish in limbo and his repeated requests to probe for oil in northern South Australia were being similarly stonewalled.

A Russian hydrogeologist in the department was conducting a chemical analysis on the waters of the Great Artesian Basin and in the course of this work, he and Reg had been reviewing regional bore records.

'You know in Russia we'd by very interested in some of these waters petroleum-wise,' he told Reg. 'We can tell whether there's been oil present or whether the oils are migrating. If I was looking for oil in this state, I'd be looking at the north-east corner,' he said.

Reg was intrigued with his comments, especially as it seemed to dovetail with his own finding of natural gas escaping from some of the bores. He urged Dickinson to allow him to map the western extremity of the Great Artesian Basin for its hydrocarbon potential. But was outright denied.

Dickinson was vehemently opposed to there being any oil in South Australia. His predecessor, government geologist LK Ward, had already discounted the state's oil prospects—those even of the whole country—and

Dickinson maintained the same gloomy opinion. He adhered to Ward's advice to him—'don't waste too much money on looking for oil'—and focused his attention on uranium as a much more viable energy source.

More and more Reg doubted his future in the department. There had been such freedom in the early days and he hated to be hamstrung by politics and personalities. He had more than fulfilled his commitment to the government—made prior to his overseas trip in 1948—to remain in its employ for at least three years after his return, and he felt that the time was now ripe to consider new horizons. Griselda hated to see her husband frustrated by the work that normally yielded him so much pleasure. She agreed that perhaps it was time to move on.

As REG STOOD pondering his future, awaiting a tram car on King William Street, he was startled from his reverie by a newsboy's chorus: 'Oil gusher in WA! Oil Gusher!' Ampol, in cooperation with its West Australian Petroleum (WAPET) exploration partner, Caltex, had struck oil at Rough Range on the very western edge of Western Australia, on the shores of the Exmouth Gulf.

The papers were full of news of the discovery—pictures of a grinning geologist smeared in black goop attempting to hold a bucket in the high pressure stream; and there was Ampol's chairman Bill Walkley strolling along Pitt Street, Sydney, in a blazing red ten-gallon hat, fulfilling a promise he'd made to journalists in the event of an oil strike. Reporters described the find as the biggest event in Australia's history since the discovery of gold and prime minister Menzies confirmed that it was a 'very important development' that may have a tremendous bearing on Australia's internal economy and international position.

Reg's skin tingled with a wild excitement. It was 4 December 1953 and dry, old, continental Australia had struck its first ever 'flow oil.' Like hundreds of others around the country Reg rushed to buy Ampol shares, pushing the price up from 10 shillings ($1) to £6 15s ($13.50) in a single day and heralding the beginning of an oil boom.

Within days he had a visit from a Robert Bristowe 'a flabby, red-faced individual, late of the "diplomatic service",' who wished to discuss oil prospects in the Lake Torrens area north of Port Augusta. Bristowe claimed to be backed by the prestigious Bonython family—owners of Adelaide's *The Advertiser* newspaper, among other things—and said if necessary he could raise £1.5 million to set up a large South Australian oil exploration company.

Geologists were in demand like never before and throughout the next month Reg was inundated with calls from mining companies and prospecting syndicates with requests to join their exploration programs. His old friend Maurie Mawby said he had never understood Reg's decision to remain in government service rather than join Zinc Corporation when he had offered him a position back in 1950. He impressed on Reg that now was the time—if ever there was one—to leap into private enterprise. Oil and uranium were the toast of the stock market and Reg was an expert in both. For a geologist of his credibility and experience there were options aplenty.

Instead of accepting another position, however, Reg favoured setting up his own outfit. He could consult for whomever he chose and have more control over how he would divide his time and where he would explore. The thought of being able to conduct his own explorations unfettered by departmental bias was irresistibly tempting.

The general excitement was contagious and soon others in the department were also itching to somehow get involved. Alick Whittle and Bruno Campana were keen to form a business with Reg, and Bruce Wilson had asked to join if the company went ahead, even offering to work for a year without pay to get things started.

The more Reg considered the idea, the better it seemed. The proposed Bonython show was anxious to formalise an arrangement—anything from a directorship on the board to appointing Reg as John Bonython's personal geologist with a guaranteed retainer and the freedom to seek other consulting jobs, was on offer. A local uranium mining promoter was also keen to engage his services, and others from Zinc Corporation to the

Commonwealth's Bureau of Mineral Resources were promising to push work his way if his new company went ahead.

Reg worked late into the nights preparing costings. He had already decided on a name for his new organization: Geosurveys. But there were so many other details to resolve. Word had already leaked out of 'Sprigg's new venture' such that his departure from the Department of Mines was virtually a fait accompli. Geologists at the university such as Arthur Alderman (who had been appointed the new Chair upon Mawson's retirement) and Eric Rudd professed to be delighted at his move but also expressed concern at the depletion of talent in the government service. 'The mines department will never be the same!' exclaimed Hector Brock, the geology school's long-standing technical assistant, who, with Mawson, had shepherded Reg through his student years and viewed his career with interest.

Before he was ready to submit his official resignation however, Dickinson phoned him at home, in the middle of an evening dinner party, to unleash a tirade. He swung from threatening to cajoling; from intimidation to abuse. He accused Reg of unproductivity, poor workmanship and unreliability, all of which Reg robustly repudiated. It was an unnerving and upsetting 45 minute blast, which finished with Reg submitting his resignation on the spot and with Dickinson predicting failure and saying how silly Reg was to get mixed up with these 'fly-by-night' companies that would all be broke in 18 months anyway.

It just made Reg more determined to go ahead. He submitted his written resignation on 26 February and turned his eyes firmly to the future.

On 1 March, at 3.40 am the ground shook. It was South Australia's worst recorded earthquake. The rumbling lasted 20 seconds and registered 5.4 on the Richter scale. Reg awoke as the house shook and raced to grab his crying daughter. Back under the covers he lay visualising the quake's epicentre 12 kilometres south of Adelaide and seeing in his mind's eye rocks slipping along the Eden Fault line.

It was his thirty-fifth birthday; the earth was moving; and Reg's life was in for one hell of a shakeup!

Dickinsonia lissa

CHAPTER 6

OIL

REG AND GRISELDA rose from their bed on Monday 1 March 1954 and hastened to the bank. After being debt-free for the first three years of their marriage they were now about to embark down the path of mortgages, loans and overdrafts. First up was the purchase of a Landrover for Geosurveys fieldwork. Within a month they had mortgaged their home. But there was excitement in the air and Reg felt that prospects were good. Even Griselda's announcement that she was expecting their second child did not see him falter. Although Dickinson had now softened and was promising his old job back if things turned sour, Reg knew that there would be no retreat. He was ready for a challenge and full of enthusiasm for what lay ahead.

On the 11 March his private company was incorporated with just he and accountant Keith Cambrell as subscribers. An accountant at the mines department had pressed on him the importance of having someone dedicated to running the books and watching expenditure, and Reg had heeded the advice.

'Geosurveys of Australia has been formed in Adelaide to assist public companies engaged in the development of Australia's mineral and petroleum resources,' Saturday's Adelaide *Advertiser* announced. 'It will undertake on a fee basis consulting and contracting work, and generally provide information and specialist advice on geological and related problems. Nominal capital is £20 000.'

The new week started with Geosurveys truly launched. 'Glorious to be free of Government red tape and Master of my own destiny,' Reg wrote elatedly in his diary. 'No longer to subjugate brains and professional ability to a bureaucratic machine.'

By now, Bonython's proposed company, too, had a name: South Australia Northern Territory Oil Search with the acronym Santos. The fortunes of the two companies appeared set to be linked. After Reg and Cambrell had their first official Geosurveys Board meeting Reg joined Bonython, Mawson, and Bristowe to discuss formation of the Santos Board. In the afternoon a uranium mining promoter arrived with a cheque for £1000 and a contract letter for £4000 per annum for services as rendered. From his office in the Steamship Building in Currie Street, Reg then drafted an advertisement for four more bright and ambitious geologists to join his team. He was offering £2500 per annum for senior geologists—only £250 less than Dickinson was earning to head the mines department; £2000 for qualified geologists—at least £300 more than his former annual wage; and £1500 for assistants. Wages for geologists in private enterprise had escalated as mining fever took hold and Reg's small company had to keep pace to attract the best candidates.

The lucrative offers alarmed Dickinson who moved to stem the flow of qualified men from his department by boosting the field allowance for his staff.

'I wish to request urgent attention to this matter as the position has become most critical,' he wrote to the Chairman of the Public Service Board, who held the purse strings. 'The attached list indicates the officers who have resigned from the Survey over the past 12 months—it totals 11 out of a field strength of 29, and the drift is still continuing. Included in this list are such outstanding geologists as Messrs Sprigg and Miles, together with a number of other experienced and competent officers whose replacement will be exceedingly difficult. It is essential to offer every possible inducement to attract competent men, besides trying to retain the present staff, if the department is to carry out its commitments, and maintain the necessary standards for geological work.'

Even the media commented that 'Geologists in South Australia are becoming almost as valuable as the uranium they seek,' and noted how Reg and other experienced men were leaving government service for the excitement and opportunity offered by private mining and exploration companies.

Reg found the whole climate invigorating. The long hours in meetings and the financial uncertainties were gruelling, but he was fuelled by the knowledge that his first pay cheque was in the bank and enquiries for his services were rolling in. The next day he was off to Mount Gambier to discuss the Otway Basin with experts from international companies. The following day he would fly to Perth. The state was still euphoric over the Rough Range discovery and he had various offers of work there to consider. Over lunch he mentioned his recent sojourn in the Tomkinson Ranges and the evidence of nickel his party had found there but had not had time to adequately investigate. A spark of interest flared in his dining companion's eyes and he immediately asked Reg to write a report on his findings.

By the time he returned from Western Australia, Santos had been incorporated and there was a pile of job applications on his desk. With demand for geologists so high it was reassuring that so many—including a professor and a consultant to a major company—were attracted to his own show. He should be able to identify those he wished to hire from among them. He was particularly interested in a Swiss geologist, Dr Rudi Brunnschweiler, who had flown his Sokol light aeroplane to Australia five years before. He was a geologist and a pilot with a sense of adventure—a combination that had immediately appealed to Reg when he had met Rudi recently at his home in Canberra.

Reg spent early April writing the foreword for the Santos prospectus, as well as a simplified geological write-up of the oil prospects of the Lake Eyre Basin. He also considered the physical location of his own business. An office in town was convenient but he also needed rooms to house the staff he anticipated appointing, as well as storage areas for field equipment and vehicles. Perhaps he could rent some space from his father.

Pop Sprigg's Woodville property was large—occupying the equivalent of eight house blocks—and he lived there on his own since Reg's mother had died the year before. When Reg suggested renting some rooms in the house Pop was most enthusiastic and readily approved the minor reconstruction necessary. He also wished to join the roster for field assignments. Hell, he was only 68 and had a few good years of tramping the country left in him yet, he assured his son.

True to his word, three months later Pop was camped south of Adelaide River, 100 kilometres south of Darwin, sniffing around for uranium near to where three prospectors had stumbled upon a deposit during a hunting trip a year before. Reg had assessed the prospectors' find on behalf of Uranium Development and Prospecting and recommended that the company purchase the lease. He had immediately recognised spatterings of brilliant grass-green torbernite and he suspected an underground deposit of pitchblende—uranium ore in its purest form. Within days Pop had discovered another rich torbernite outcrop.

The timing was propitious on several fronts. Uranium Development and Prospecting could supply ore to the uranium processing plant due to open at nearby Rum Jungle in September. More importantly, payments to Geosurveys for development of the Adelaide River uranium mine would keep Geosurveys on a stable financial keel. While Reg was steadily increasing his payroll and expanding his fleet of vehicles and equipment, expected income from Santos was receding into the distance. The directors were still dithering about when to publicly float the company. Since the market had cooled they felt it better to wait for Ampol to make a new oil strike and ride on the back of the anticipated upsurge, than dive into the financial fray on their own.

In June Rudi arrived to begin working for Geosurveys and together he and Reg worked on the Santos prospectus in readiness for the public float. The budget allocated £25 000 for exploration work in the Lake Eyre Basin, £8000 for Wilkatana investigations and £24 000 for the Northern Territory and Queensland's Bowen Basin, west of Rockhampton. Rumours were flying that Ampol/WAPET's Rough Range-2 well had struck oil at 3015 metres. Hubert Harvey felt sure that Santos would now 'go ahead like wildfire' when they floated in two to three weeks. Speculation on Rough Range's prospects was running wild again but Reg knew that it was too early to judge success or failure.

In July Reg went with Bonython and Bristowe to inspect the Wilkatana bore that was to be the jewel in Santos's crown. It was here on these sparse plains in 1934 that a driller had reported oil rising in a water bore. When former Government Geologist Dr LK Ward had examined the sample as part of his statewide assessment of oil showings, published in 1944, he had grudgingly labelled the Wilkatana oil show the 'least intangible' of the lot. Only 1.2 grams of oil were delivered to the Department of Mines for analysis and although it was pronounced genuine, nothing further was done.

Bonython's family owned Myrtle Springs Station a few miles further north and young John had never forgotten his great-uncle's prediction that oil would one day be found in these parts. When John's former school-mate, Bob Bristowe, returned from the war they had renewed acquaintance and Bob told John of his travels with the British Foreign Office as King's messenger—hand-delivering diplomatic dispatches and flying back and forth over the deserts of the Middle East where oil now flowed in such exuberant profusion. The terrain was a lot like that of outback South Australia he remarked; dry and brown and barren. The two men's thoughts coalesced and the kernel of Santos was formed right there. They embarked on a reading and research programme at the public library to find out all they could. And then Rough Range happened. It was time to put old Alfred Bonython's convictions to the test.

Near a stand of myall trees at the western foot of the Flinders Ranges, Reg, Bob and John gathered around the old Wilkatana bore pipe and watched as minute globules of oil flowed out with the artesian water—oil that had been leaking out for twenty years and that Dr Ward had judged so casually. For John Bonython, seeing the oil rise to the surface was one of the most exciting days of his life. The station manager then led them 100 yards away to another bore drilled in 1946, which had never been officially examined. Here too, Reg was delighted to find globules of oil rising to the surface as they stirred the water. He looked carefully for evidence of a 'plant' but none was forthcoming. There was authentic oil here all right. It was just a question of how much.

Preparations to float Santos recommenced. Reg was now juggling three major projects: Geosurveys had commenced a regional survey north from Port Augusta and Reg was organising aerial photography of the Wilkatana area courtesy of good friend Allan Vial. He was also making regular trips to the Northern Territory for the uranium find and, furthermore, investigations were about to begin in the far north-west of South Australia. Reg had joined with others to form Nickel Mines of Australia to investigate the nickel occurrence at Mount Davies.

On the 8 October—finally—Santos was floated as a public company, issuing 1.5 million shares at 2s 6d each, and putting £187 500 in the kitty. Chairman of the board, was John Langdon Bonython, solicitor and director of several public companies. Bob Bristowe, and retirees surgeon Sir Henry Newland and Major General George Symes, were company directors. In September Mawson had resigned his directorship due to illness, but enjoyed the new title of Honorary Consultant. Geological consultants were Glaessner, Sprigg and Brunnschweiler.

The names of those associated with the company had considerable cachet in Adelaide society circles and Bonython announced the float to be a 'substantial success.' In addition to the 300 000 square kilometres of the western portion of the Great Artesian Basin and the Lake Torrens Basin in South Australia, Santos now held oil exploration licences over 13 000 square kilometres of Queensland's Bowen Basin, and applications were pending for

OIL

additional areas in the Northern Territory. It was a huge acreage and never again would one company be awarded exclusive control of so much land.

By the mid-1950s much was already known about the structural requirements for oil accumulation in commercial quantities. Firstly there must be reservoir rocks—solid but porous sandstone beds capable of retaining oil in much the same way as a brick soaks up and retains water. The oil beds must then be covered by non-porous layers of clays or shales that create an impervious cap. Lastly there must be some kind of subterranean rock folding, often leading to a telltale dome or anticline on the surface, which traps and thus concentrates the oil in certain areas. This much was known.

Now Reg used the Santos prospectus to at last put forth his theories on the specific oil prospects of the Great Artesian Basin—where he had wanted to explore for so long, and where he so passionately believed oil to lie.

'The Great Australian Artesian Basin, in which the Lake Eyre concession lies, is one of the major sedimentary basins of the world,' he wrote. 'The basin has been the site of large inland seas, in particular during the Cretaceous Period, at which time many of the deposits containing the larger oil fields of the earth were laid down. It is an area of thick sedimentation which is known reasonably well only from shallow borings, and is consequently in urgent need of reassessment for its oil possibilities, particularly in the light of recent advances in geological thinking on the origin and accumulation of petroleum…the presence of oil in these great basin areas should be expected.'

He described how ancient shifting coastlines around marginal salt lakes, estuaries and swamps provided very favourable environments for the generation of oil. Buried organic matter from the Cretaceous Period 65 to 145 million years ago now formed blue-black beds of marine sediments at least 600 metres thick. These sandy and shelly beds were draped over bedrock irregularities that provided excellent structural and stratigraphic traps for migrating oil. Furthermore the beds were capped by as much as 500 metres or more of carbonaceous shales to retain oil migrating from the underlying marine sediments.

He pointed out that drilling into these beds in Queensland at Roma and Longreach had already demonstrated the presence of petroleum and 'wet' gas in a numbers of areas. Cretaceous marine sediments were some of the most prolific producers of the world's petroleum and it was in these types of sediment beds that the Rough Range discovery had just been made. He explained how the water that supplied the bores of the Great Artesian Basin was restricted to beds well below the known marine horizons and dismissed the claim that no oil in water bores meant no oil was present. Experience elsewhere in the world had already shown that water bores had frequently failed to reveal what later proved to be important oil fields. Furthermore there were still great zones of at least 50 000 square kilometres where not a single bore had been drilled.

He also highlighted that nothing was known of the sediments that lay much further underground and hinted that here, much older Permian layers from 250 to 300 million years ago, deep in the basin's bottom, may contain surprising oil reservoirs in ancient glacial sediments. He did briefly mention the oil shows in the Wilkatana bores, but it was clear where his loyalty lay.

'In view of the foregoing undoubtedly favourable features for the generation and accumulation of petroleum it is obvious that the Lake Eyre and related sedimentary basins are thoroughly worthy of the fullest geological and geophysical exploration followed by test drilling. A strong recommendation is hereby made to this effect,' he concluded.

REG SPENT MUCH of the next two months interstate: Queensland, the Northern Territory and Western Australia. When he *was* in South Australia he was usually away on field assignments. His transition from public servant to private consultant had certainly not created more time for family life—far from it. Griselda and Margaret were largely left to fend for themselves while Reg blew in and out like a tornado, bringing rolled-up maps and fat files; love, good cheer and dirty laundry; and departing like a vacuum that sucked the happy bluster out of the house in the black morning hours. Griselda was stoic and uncomplaining. She kept house and typed her husband's reports and told two-year-old Margaret about her daddy's important exploits.

OIL

And soon there would be another little body clamouring for attention. At Christmas time the new baby was due. She and Reg both hoped it would be a time when the growing family could snatch a few days together.

When he left for Perth towards the end of November Reg kissed his wife goodbye and promised to be back as soon as he could. When he next saw her she was lying wan and anaemic in hospital and he had a new son. Douglas Paterson Sprigg, named for Reg's revered Professor Mawson, had arrived five weeks early in the middle of a Geosurveys drama. Like his mother and sister before him, he would soon learn that family was interwoven with work and that geology was not a career but a way of life.

Geosurveys's first prospecting party, comprising old hands Syd Giles and Smiler Greenwood—both in their sixties—and young enthusiast Dennis Walter, had been camped out at Mount Davies since the beginning of September, investigating Reg's north-west nickel find. When word came through on the pedal radio that they were almost out of water it was up to Griselda to contact Todmorden Station 80 kilometres north-west of Oodnadatta, for help in organising a mercy water delivery. By the time she received word that the water-carrier was on his way, so was her baby. She left Margaret with the neighbours, pinned a note to the door, wedged herself into the spare Geosurveys Landrover, and drove herself to hospital.

It was just like last time, when Reg had been away when Margaret was born. Not that she really wanted him in attendance throughout. Griselda had worked in enough maternity wards in her nursing days, to know that, for her, the actual delivery was one time when she'd rather be on her own.

Reg arrived home at 6.00 am the next morning and was brought up to speed with events both at Mount Davies and at Calvary hospital. He then had news of his own to share with his wife: He'd be off next morning to Mount Davies, himself. The world leader of nickel, International Nickel Company, had sent representatives out from Canada to assess the new Australian find. The chief exploration geologist was anxious to see the deposit for himself and wanted Reg to show him around. Less than 24 hours later he was gone again.

It was another five days before Reg got to meet his son. He returned from the field full of remorse for the stress Griselda had had to bear on her own.

'Arrived back in Adelaide conscience-stricken over time away from Gris and children,' he recorded in his diary. 'Find that daughter Marg is suffering from my frequent and sudden overnight departures and long periods away. I have been away 75% of time in the last exhausting six months.'

Determined to make amends he dressed his daughter in her prettiest pink organdie frock and took her to meet her new brother. Griselda smiled when her husband entered her hospital room, and smiled even wider when she saw Margaret... wearing her best dress...back to front. Still it was Reg's first attempt at dressing his offspring and he hadn't made too bad a fist of it, she laughed.

THE NEW YEAR was devoted to preparations for a drilling programme at Wilkatana. Despite all Reg's talk of the Great Artesian Basin, for Bonython and Bristowe the oil they could see with their own eyes coming out of the borehole near Port Augusta held the most immediate appeal. The stock market was jittery following the failure of WAPET'S sixth well at Rough Range and South Australia's Mines Chief was still broadcasting his opinion that there was no oil to be found in Australia. Reg and the Santos men believed it was an attempt to discredit their operations. Only petroleum proof would quiet him, and they dedicated themselves to that end.

At the end of January 1955, the first Geosurveys team departed Adelaide for Wilkatana. Visits to the drill site 50 kilometres north of the township of Port Augusta now became a regular monthly, and often fortnightly, occurrence.

In March, Geosurveys turned one year old and baby Douglas three months. Reg submitted his first annual account to Santos and was irked when Bonython baulked at the sum of £8000 for the use of three geologists, several vehicles and a plane. Furthermore, Reg had only charged for 25% of his own time although the hours he spent on Santos affairs far exceeded that.

'Feel like pulling out. Only my loyalty to the project and my belief in the prosperity of the areas keeps me on,' he noted in his diary. He resented the company's tight-fisted attitude, especially if, as he feared, manipulating the share market sometimes took precedence over fieldwork.

It had been another hot summer and at their home in Somerton Park, Griselda was proud of the way her children were coping with the heat.

'For about three to four weeks now temperatures have been 90 to 100 °F and humidity about 86%,' she recorded in Douglas's baby book. 'Nights have been shocking too, but both Douglas and Margaret take this in their stride. Never before have I seen beads of perspiration on a young baby's brow but I have certainly for last few days. It is amazing how well he takes this heat. Being thoroughly bad-tempered about the weather myself I think both my children deserve full marks for good behaviour.'

Their tolerance for the trying conditions bode well for their future roles as junior Sprigg explorers—following their father into the extreme Australian outback. Griselda knew that the only way that the family could be together was if she and the children became mobile 'camp-followers.' Reg had missed so much of his daughter's early years and Griselda was determined not to be housebound forever. If Reg wouldn't come to them, she decided, they would go to him. The children must be readied as soon as possible. Most important was to dispense with the baby's cloth nappies and all the associated soaking and scrubbing. She had begun Douglas's toilet training in January, when he was only six weeks old.

'He objected strongly at first,' she recorded, but after a couple of days, 'now is quite content to cooperate.'

If it was unconventional to start toilet training so early, Griselda didn't care. By the end of the month he was, 'doing well with his training, although he always protests as a routine practice when first "held out".' By the middle of February he had not had a soiled nappy for more than a week.

'He uses his pot after each feed and sometimes before for tinkles, and after early morning feed he does his (very) big job with great energy in his pot. This certainly makes life very easy.' Her schedule was on track for she and the children to accompany Reg to Broken Hill at the beginning of

May. It would be a welcome chance to catch up with a few friends from the Radium Hill days.

The Radium Hill mine had officially opened six months earlier and the little site where Reg and Griselda had parked their small caravan and begun their married lives together on the lower slope of Tank Hill, now overlooked a steadily growing township of more than 500 people. One hundred and ten timber-framed houses lined The Avenue and First through Fifth Streets, and blocks of two-man cubicles comprising the single men's quarters clustered along the road to the industrial area a couple of kilometres away. In the hierarchical tradition of mining towns, professional staff and section heads lived in a secluded area, with the General Manager occupying a brick house at the top of a rise with a commanding view over his domain.

The town now had its own post office and newspaper—of sorts—*The Radioactivity News Bulletin*, produced by the community club and distributed free to every home. A revegetation and tree-planting programme was in place to reduce dust around the town and gardening competitions were vigorously contested. Mrs Pengilly, wife of the local garage proprietor had conjured a lush square of green lawn from the hard orange earth and surrounded it with a profusion of roses and dahlias that was the envy of her neighbours.

Alec Pengilly, in addition to running the service station and motor garage, also provided refuelling and ground operations for Guinea Airways, who offered twice-weekly flights to Adelaide for five dollars. On one of Playford's frequent visits to the site, Alex had been engaged to transport luggage to the plane for the group of seven VIPs.

As all the passengers milled around awaiting departure, Alec hurriedly re-positioned his truck and… 'Bang! I looked behind and there was the thunderbox knocked over,' he recounts in *We were Radium Hill*. 'I got out, and someone said there was someone in it. Someone else said there was a passenger in it, so things were getting worse. Finally someone said Sir Thomas Playford was in it!

'The [General Manager] rushed up. Sir Thomas was flat on his back and lying on the door. The GM said "We can't roll it over with him in it," in a stage whisper. "We'll have to get him out the bottom." Which we did. We dusted him off—as he only got dust on him, fortunately. We apologized profusely to him; and he said it was an accident, shook hands with me, boarded the plane and away he went.'

'It was so funny because Tom was so unruffled,' recalls Betty Rodgers, the manager's wife, who was at the airport to farewell the dignitaries. 'I nearly collapsed. He just climbed out through the hole in the bottom of the toilet, straightened his hat, walked over to Terry, shook his hand, shook my hand and off onto the plane. He didn't blink an eye and didn't say a word about it, just shook our hands and said goodbye.'

Naturally this luscious tidbit appeared in the next day's newsletter under the heading 'Upset in Parliament House.'

Social activities in the town included membership of the Radium Hill cricket or soccer team, where players competed against residents of Olary and Cockburn in ball skills and gravel rash on the town 'oval.' The rifle club and RSL sub-branch presented less painful options, however hazards for golfers on the Radium Hill course included ball-thieving crows and stone-strewn, sandy fairways where having a ball rebound from a rock and end up behind your position of play was not unheard of and searching for the ball in the scrub took up most of the game time. Other recreational outlets included an open-air picture theatre and the wet canteen—with a strictly enforced ladies bar reserved for 'ladies only and ladies accompanied by a male escort'—where beer, stout and bottled soft drinks were served. The sale of wines and spirits was considered 'most undesirable.'

Despite the relative privations and a labour turnover of up to 165%, for many residents—almost half of them migrants from up to 15 different countries—these were happy times. But there was no romance to life in Radium Hill for State Bank branch manager Hugh Dalby.

'During my time the main Broken Hill road was dirt and there were no bitumen roads in Radium Hill,' he recalls in *We were Radium Hill*. 'To me the place was dry, dirty and dusty. We had no garden only a solitary pepper tree. During the summer, the water was too hot to shower until late at night, for the water came through pipes lying on the surface from Teasdale's dam.

During winter, in the house, we opened the door of the oven and turned on all the elements to warm the joint up.

'As Manager of the early social club in 1952-53 one of the biggest problems was the supply of beer. Kegs in those days were made of wood and when left on the railway station at Olary, in the heat, they would often boil in the sun, resulting in 25% evaporation.'

The town now had a school attended by 100 children from kindergarten age to grade nine; a retail store selling groceries, confectionery, chemists supplies, newspapers and tobacco; a small timber and galvanized iron State Bank building that would contain in the vicinity of £20 000 on payday eve; and an Australian Inland Mission hospital staffed by two nursing sisters assisted by the Royal Flying Doctor.

The *Atomic Comet* pulled in to the Radium Hill railway station to dispense supplies and passengers, and ferry trucks of ore concentrate to the chemical treatment plant at Port Pirie for uranium extraction. A reticulated water system delivered water from Umberumberka reservoir, 85 kilometres away near Broken Hill, and a high transmission power line from Morgan, 210 kilometres away on the River Murray, supplied electricity to the town and industrial site. All of these utilities were completed in record time and at government expense. Development costs would gradually be recouped over the seven years following the mine's opening in the sales agreement struck between the Commonwealth and South Australian governments and the governments of the United Kingdom and the United States of America. The UK and USA Combined Development Agency also guaranteed to buy the entire uranium output of the mine until the end of the contract in December 1961. It was a handsome deal for South Australia and Playford was proud of the terms he had negotiated for his state.

The main shaft of the mine itself was down to 200 metres with provisions for deepening to 600 metres to reach the deepest ore body. Thirteen kilometres of rock tunnels had been cut on three levels beneath the ground. The metallurgical concentrator, using a magnetic method of separating the uranium-bearing minerals from the waste rock, had grown from a pilot plant with a capacity of 20 tons per day to a full-scale milling and extraction plant turning 400 to 500 tons of crushed rock into 40 tons of uranium concentrated to 1 to 1.5% every day. At Port Pirie the uranium was leached from the ore concentrate by boiling sulphuric acid to eventually produce a high-grade yellow uranium salt. This yellowcake was finally dispensed into 44-gallon drums that were welded shut and shipped to the USA.

'This is a great day. A great day for South Australia, for the Australian nation and for the Free World,' the Governor-General Sir William Joseph Slim had proclaimed at the mine's opening ceremony on 10 November 1954. 'For your State, because it marks a stage nearer the goal to which with faith and courage you are advancing. For Australia, because it is another step in her steady march towards greatness. For the Free World, because it strengthens the hands of those who would defend freedom…Uranium, the basis of atomic power, has two aspects—war and peace. To produce atomic power here in SA would be an achievement unsurpassed in Australian history—and you are on the way to it. You will bring this project to success and in doing it you will confer vast blessings on your people and your country.'

The Governor-General and his speech successor, Premier Playford, made much of Playford's hope that an experimental nuclear reactor would be built at Radium Hill within four years to produce energy to power the state. The international nuclear arms race was downplayed and the spectre of cheap, plentiful power to replace the state's reliance on imported coal and oil was dispensed to the populace with ready acceptance.

To conclude the ceremony Sir William was presented with a silver cigarette case made by the Radium Hill miners in the form of an underground ore truck, before pressing a series of buttons to officially start operations. Playford then slung a Geiger counter over his shoulder and he and the Governor-General, in overalls and lantern helmets, waded through slush 100 metres underground examining the ore lines and chatting to bearded miners, their voices frequently drowned by the roar of pneumatic drills and the thunder of distant explosions.

REG HAD ACCEPTED an invitation to attend Radium Hill's official opening but his interest was mostly academic by now. Despite being named one of three 'Men of the Week' by the Adelaide *Advertiser*—along with Dickinson and mine manager Rodgers—for being a 'key figure in the great uranium project,' his personal connection to the mining field he felt he had almost single-handedly rejuvenated, had been severed, and he could not help some lingering disappointment at the way his contribution seemed to have been largely overlooked by those in the mines department. He was not made to feel especially welcome at the festivities. Never mind, his thoughts were now on other matters.

On 3 June 1955, the Santos Wilkatana-1 well was spudded. As was standard practice the contractors used a percussion drill to pound through the earth as quickly as possible and reach the depths of interest. Within a fortnight, at 70 metres, it was flowing artesian water carrying minor globules of oil. Reg was still interspersing visits to Wilkatana with trips to the Northern Territory uranium fields. He was also using his personal

Griselda nurses baby Douglas at the Wilkatana field, 50 kilometres north of Port Augusta, site of Santos's first oil well, 1955.

contacts in the Commonwealth Bureau of Mineral Resources to arrange a seismic survey of the Great Artesian Basin and aerial photography of the Wilkatana area.

At the end of November Reg had discussions in Melbourne with Zinc Corporation Chairman Maurie Mawby. A Zinc Corporation offshoot called Frome Broken Hill Company (Frobilco) currently held the exploration licence over the north-east corner of South Australia—the part of the Great Artesian Basin that Reg most wanted to explore. Over the past nine years the company had carried out a detailed gravity survey and drilled several test wells but found nothing to encourage their search except for some traces of dry gas and a gravity anomaly around Innamincka that was never followed up. Reg knew that interest in the area was waning. Perhaps Frobilco would abandon their exploration activities there and he could secure the licence for Santos. Mawby promised him that if they decided to relinquish the acreage, Reg would be the first to know.

Four days before Christmas, Bonython phoned to report heavy grease 'flowings' in the Wilkatana bore and on Christmas Eve the company announced oil at 175 metres. The day after Christmas Reg went to Wilkatana to inspect the oil 'show' and was disappointed with what he saw. Nothing was actually *flowing* out of the well. It looked to him like the remnants of ancient residual oil. Reg had always believed that the oil accompanying the artesian water in the Wilkatana bore came from 500-million-year-old Cambrian rocks not the more prospective younger rock layers, and the look of this sludgy discharge seemed to confirm his opinion. Unless it was contamination from the drilling rig itself, or even...the unwelcome thought of well salting—an old mining term for when favourable samples are deliberately planted in a mine to fool observers—crept into his mind. If so, it was a rather clumsy attempt. This heavy, Vaseline-like grease was nothing like the fine globules of light golden oil that he and others had seen rising spontaneously from the artesian water bores.

Reg's suspicions grew, however, when an official announcement to the press of the 'oil strike' at Wilkatana sent Santos's shares soaring from less than 6 up to 22 shillings.

'Reporters etc converge on "oil bore",' he recorded in his diary. 'Gob of grease conveniently pulled with bailer. Am still uncertain as to what it all means, this "grease". Santos shares in Adelaide rose from 6 to 15 shillings: worrying.'

On the 28 December 1955, Sir Lyell McEwin, the Minister for Mines in South Australia, announced that Santos had been granted the oil exploration rights for the north-east corner of South Australia. The company now had exclusive control over more than 400 000 square kilometres of Australian countryside. Mawby had kept his promise and phoned Reg as soon as Frobilco's acreage was officially surrendered. Reg had immediately prepared Santos's application and submitted it to the Department of Mines the next working day.

'Obviously the area is no good for oil,' the registrar had told Reg upon receiving the application, 'Santos may as well have the lot to keep the other sharks out and save the department more trouble!'

In May, Geosurveys's new geologist, Dr Helmut Wopfner arrived. 'Heli' as he universally came to be known, was a 31-year-old Austrian fleeing the constrictions of postwar Europe. Formally studying the childhood geology that he loved had eased his turmoil brought of being an ex-German-Air-Force prisoner of war, and now, with a PhD under his belt, he was ready for new horizons. Through the international geological community he had made contact with fellow 'flying geologist,' Rudi Brunnschweiler, who assured him that Geosurveys could put his skills to good use.

Rudi was at the docks to meet him when he, his wife, Inge, and their 10-month-old son, Klaus, arrived in Adelaide. Rudi then drove them to the Glen Osmond Golf Club in the Adelaide foothills for a view over the city. It helped soothed the initial shock Heli had felt as he had stared with dismay at the low corrugated iron buildings and straggly trees lining the route from Outer Harbor. So different from his native Innsbruck.

Within a week he was at Wilkatana to view operations and receive a general reconnaissance of the Flinders Ranges. Touring about in a Landrover without brakes for four or five days he found disconcerting, but fun nevertheless. There was little time to help his wife settle in to new accommodations and a new country. She was to learn to adapt to the life of a field geologist just as Griselda had done. After a couple of weeks in the ranges Heli was back in Adelaide for a week and then off to Oodnadatta for a three-month mapping stint.

Reg was not immune to the imposition it would place on the family, but work was a priority, he reminded Heli's supervisor, Rudi. 'As it is essential we get a maximum of field work completed by November it means a period of some hardship for Mrs Wopfner...It would be foolish not to get this area under control before the hot weather.'

It was now mid-1956. The Wilkatana-1 well was down to 670 metres and Santos had spudded its thirteenth well in the Wilkatana field—a narrow belt 16 kilometres wide and about 80 kilometres long. A team of 16 men, comprised of Geosurveys geologists and Department of Mines drilling crews and a cook, were camped on site. By now Reg had put any concerns of 'salting' out of his mind. He had examined the cores from the drill holes himself and seen the yellow-green fluorescence that indicated oil in the rock. Oil analysis reports from Melbourne confirmed the new samples were genuine and he trusted the field superintendent on site.

'All now looks convincing even to my cautious mind,' he noted.

He and Bonython presented a sample of the oil to Premier Playford who was enthused by the thought of a great new energy resource in his state—where government experts had always advised against it.

However, Reg was still sceptical about the real prospects of a commercial oil field at Wilkatana. He felt these Cambrian rocks harboured a depleted petroleum reservoir from which the oil had leaked away many millions of years ago. While fulfilling Geosurveys's obligations at the site he also pursued his investigations of the Oodnadatta anticline that Rudi had mapped and quietly nourished his appetite for the Great Artesian Basin. Santos company funds were dwindling and the directors were in no mind to embark on a potential wild goose chase deep into the South Australian outback. Thanks to Reg they had the acreage, but not the money or the inclination to explore it. But Reg was in charge of his own company and staff, and he surreptitiously dispatched his geologists north to Marree,

Birdsville and Innamincka to conduct gravity surveys, and he set Bruce Wilson to collating all the geological information he could find on the wells already drilled across the Great Artesian Basin.

'I'm willing to bet that oil will be found in the Great Artesian Basin in the next decade,' he told Bruce. But there was no one to accept his wager.

In January, however, the famous American oil consultant Al Levorsen would be visiting Australia and Reg relished the opportunity to present his ideas. He had just read Levorsen's recently published textbook, *Geology of Petroleum*, and knew him to be a creative geologist with vision and an open mind. He looked forward to the American's views on the Great Artesian Basin.

In the meantime, Reg and Bonython conducted the 'Oil Prospects in South Australia' roadshow, giving joint lectures on Santos's operations to packed audiences at the Royal Geographical Society, Rotary, and the Port Augusta town hall.

By the time Levorsen arrived in Adelaide on 17 January, Reg had reports and geological plans ready for his perusal. Arville Irving Levorsen, or 'Lev' as he introduced himself, was a professor at Stanford University in California who consulted on petroleum for oil companies and foreign governments. His unconventional thinking and willingness to explore new ideas and new territory had led to his discovering the giant Fitts oil field in Oklahoma, containing 190 million barrels of oil. Now he travelled the world preaching the gospel of unorthodox thinking as the scientist's most valuable oil-finding tool.

Reg gave Lev a run-down on the general geology of Australia and could see that the prospects of the Great Artesian Basin took his fancy right away. Perhaps it had something to do with the way Reg pitched it.

'As long as there are marine sediments there,' he told Reg, 'all that is then needed is a deep section to attract US exploration money.'

To Reg's delight he recommended an exploration focus on the region.

'It's a case of Cadillacs versus Chevvies, John,' Lev explained to Bonython. 'You'll get your Fords and Chevvies around here [Wilkatana], but you'll get your Cadillacs and champagne further north.'

OIL

Since Bonython was financing Lev's trip to Australia the Santos Chairman was inclined to heed his advice. He felt a lot more confidence in the recommendations of a world-renowned American petroleum geologist who had recently visited the site of major oil discoveries in North Africa, than he did in those of his home-grown geological consultant Reg Sprigg. Yes, Reg had secured the vital exploration leases over the area on behalf of Santos, and on his own initiative, but obviously an overseas expert would know more about actually finding the oil, he reasoned. Overseas expertise was always best: Everyone knew that.

While Lev returned to the States, promising to scout for American oil companies willing to partner with Santos and bring in much-needed finances, Reg began planning an expanded field programme for the Great Artesian Basin. On a rare Sunday afternoon at home he pondered the possibilities of what might lie beneath all those thousands of kilometres of arid inland country. Recalling a survey of the region by an early assistant government geologist, R Lockhart Jack, Reg dug out the manuscript and began to read.

In 1924 Jack had roamed the drier regions of the state mapping the geology in search of shallow underground water supplies to aid pastoral settlement. As Reg re-read his report on the north-east of South Australia he felt his excitement grow and marvelled at how this pivotal information had been overlooked for so long. He, himself, had read Jack's paper long ago but failed to appreciate, until now, the significance of Jack's findings to oil exploration. And certainly no one else seemed to have picked up on it. It showed how the mind compartmentalised information and how puzzles were often solved by making the right intellectual connections. It brought to mind the words of another famous geologist, Wallace Pratt, in one of Reg's favourite quotations: 'It is the genius of a people that determines how much oil shall be reduced to possession; the presence of oil in the earth is not enough. Gold is where you find it, according to an old adage, but judging from the record of our experience, oil must be sought first of all in the minds of men.'

Reg chided himself for not seeing it sooner. Here, in this 1925 bulletin to the Department of Mines, Jack described a great dome in the desert sandstone—100 kilometres in diameter and 150 metres high. Under the surface of this anticline lay more than half a kilometre of Cretaceous sedimentary beds—the same layers that had yielded oil at Rough Range and were most propitious for oil. So there, on Cordillo Downs Station, squarely in Santos's permit area over the Great Artesian Basin, were two of the most important oil search parameters in concert—65- to 145-million-year-old sediments pushed up into an anticline. This was the place to search, Reg felt sure. If the dome was really there as Jack had described it, this would be the place to drill.

The first thing was to protect Santos's flanks and apply for additional exploration rights over the south-east Queensland portion of the Great Artesian Basin. With another 100 000 square kilometres of acreage in the Santos coffer Reg then called on his trusty aerial mapping contact at the Lands Department to request photography of the Cordillo Downs area. His old friend Allan Vial agreed to virtually smuggle a Geosurveys geologist aboard on his next north-east run to photograph flood areas along the Diamantina River and Cooper Creek. It was highly unorthodox to take a passenger on a survey flight, but Allan trusted and respected Reg, believing him to be one of the state's true patriots. He knew Reg wouldn't ask for such a favour without good reason. If it was *that* important for a geologist to see the area with his own eyes, then Allan would allow him a seat on the plane to stare out the window while his crew took their photos.

Much as Reg would have loved to see Jack's anticlines for himself he knew he couldn't be spared for the 1000-kilometre journey to the South Australian corner. He had to deputise.

Heli Wopfner remembers the day that Reg collared him at Geosurveys's Woodville offices:

'Heli, do you have time to go up to Cordillo Downs?' Reg asked.

'Where the hell is Cordillo Downs?'

'Well, it's up in the north-east corner. The Lands Department are flying the aerial photographs and I have arranged that you can do some flying with them. And I also arranged for a Landrover to be up there so you can do a little bit of the ground recce.'

'"All right," I said. "When do I leave?" This was at 11 o'clock in the morning,' Heli adds.

'Well, our secretary booked you on the Butler airflight to Broken Hill at three o'clock this afternoon!'

It was typical of Reg's impulsive style. But the Geosurveys staff always rose to the challenge. Their leader was passionate and enthusiastic and optimistic, and the geologists and field assistants, right through to the maintenance men and office workers, all felt that they were achieving something. They were part of a pioneering team, where every day was a new adventure and, 'I wonder what new idea Reg will have for us today?' was a source of endless speculation.

'There was so much going on,' Betty Wilson (nee Ewers) remembers fondly. She joined Geosurveys in 1955 and married geologist and fellow employee Bruce Wilson in 1966.

'Reg'd get so *enthused* about something or other, and after a period that would fade out and he'd get enthused about something else. He was always doing something different.

'It was a wonderful place to work. For me, it was a new life. I'd only been a typist for the council—just an ordinary sort of a life. When I went there it was like a new world opening.

'I got the job because I could read Reg's handwriting. I had to type something that he'd written—that was my test. He used to write quicker than you could do it in shorthand. That's why you could never read it.

'But that was Reg. He got everything done very quickly. He'd be here, there and everywhere; often away. We always called him the "late" Mr Sprigg. He was never where he should've been on time; always rushing in and rushing out. He left all his gear piled up at the front door—he had to put it there otherwise he'd forget everything—and he'd gather it all up as he went out.

'The boys were all coming and going. I used to envy them, always going away. But then after I married Bruce I realised that going away is not such a lot of fun. They moved around a lot. They were away for months out in the bush.

'I was there to type reports. I used to do 11 carbon copies on a manual typewriter. You couldn't read the last paper, almost. The paper was very flimsy.

'I'd do whatever they gave me to do. Specimen Minerals put out a magazine and I typed all that. Another thing that Reg had was a compilation of all the exploration stuff that was going on, the scouting service. And I did all that. I used to go down with Dennis and make up prospectors' kits for students. I did all sorts. I loved the job.

'Pop Sprigg was still living in the house,' Betty continues. 'Geos had everything else but Pop's bedroom. He wasn't involved in the business—he'd just potter around. He'd come and talk to us, or we'd go and talk to him—wondering why he wasn't up, if it was that time of the morning; checking if he was all right. He had a big garden, with a big mulberry tree out the back and 12-foot holes covered over with bits of galvanized iron that Reg had dug as a kid, as a budding geologist.'

Betty remembers Reg struggling with the finances in the early Geosurveys days—juggling accounts to keep the growing new business afloat. But the over-riding memory is of a dynamic team with a strong sense of purpose and camaraderie.

'People had this pioneering feeling,' says former Geosurveys geologist, Bob Laws, who joined the company straight from university early in 1962. 'I spoke to Reg for about two minutes and he hired me. Reg was always one for making his mind up quickly.'

When 20-year-old Bob travelled to Adelaide to begin his appointment it was the first time he had been out of New South Wales. A short time later he was camped out by the remote Northern Territory-Queensland border. Much of his eight years with Geosurveys was spent in the field:

'I remember in 1963 I spent more time under the stars than in a tent, and more time in a tent than in a caravan, and more time in a caravan than under a solid roof. But I was happy with that. They were some fantastic times. And Reg was such a personality—driving the organisation. He was a vibrant, brilliant man; sort of unstoppable. People had so much respect for him and his drive. And the pioneering nature of it all.'

WHEN HELI RETURNED from Cordillo Downs in April 1957 he came with a hand-drawn map on which he had marked, 'all these magnificent anticlines. When I came back quite a number of people would say, "Now what sort of grog had you been drinking up there that you had all these visions,"' he laughs.

But Reg had no doubts. Heli's reconnaissance confirmed Jack's observations. The Great Artesian Basin housed great oil-field-type anticlines. All the information was slowly coming together. When Reg mentioned Heli's findings to his good friend Richard Grenfell Thomas, Dick consulted his diary of a trip he, too, had made to the region in 1919 and found that he'd noted oil slicks together with petroliferous odours in the Patchawarra bore just 60 kilometres or so south of the Cordillo Downs homestead. The clues pointing to the north-eastern corner of South Australia being a potential petroleum province were disparate but convincing. It was high time for some dedicated investigations.

Meanwhile, after 24 wells drilled at Wilkatana, the field was finally abandoned. Santos's first drilling operation was not a success. The traces of oil they had found seemed to be the lingering remnants of an ancient oil pool. There was no point chasing it any further. The company continued with plans to drill wells at Oodnadatta on the western margin of the Great Artesian Basin and at Motpena, near Parachilna, but at Reg's instruction, most of the Wilkatana camp was moved to Haddon Downs, a few kilometres north of the Cordillo Downs homestead. It was time to probe the Great Artesian Basin in earnest.

Rudi and Heli were up there already, mapping a big anticline at Haddon Corner in a manner that only those two could accomplish—both being pilots they could map from the air in Rudi's little two-seater Sokol plane, now owned jointly by Geosurveys and Santos. They would take turns—one piloting, while the other plotted the altimeter readings as they flew along at low level, parallel to the surface of the outcropping structures. In this way

Reg's photo of anticlines in the South Australian corner country in the late 1950s. An oilman's dream!

they produced the first structural map of what would later become known as the Cooper/Eromanga Basin.

Rudi feared that the sediments were Devonian—much older than expected. But Reg felt sure that Jack's interpretations were correct and that the sediments were of the Cretaceous age they were looking for; and it was not long before he was proved right.

Within a couple of months, Reg finally had his first chance to see the area for himself. As Rudi flew him towards the very edge of the state he was thrilled to see a series of flat-topped hills rising up from the desert plains. In his mind's eye he saw how the top of the hills had once joined each other to form one great gentle arc. The gullies and wide valleys that now separated them were the result of eons of erosion. Small cracks in the 'duricrust' surface of the once-giant dome had slowly weathered and widened over millennia until whole sections had turned to gibbers and desert sand and skittered away, leaving isolated columns in a broken rampart to tell the story of what lay beneath.

Cordillo Downs. Haddon Corner. In this 200-kilometre square where South Australia poked into Queensland he saw the best oil structures he had ever seen. Between Innamincka in South Australia and Durham Downs in Queensland he discovered another huge dome, right where the Frome Broken Hill Company had in 1947 recorded a positive gravity anomaly—an indicator of possible oil. Eight years later they relinquished the lease and Reg snapped it up for Santos. If only they could now see what he could see! A great anticline 80 kilometres long and 30 kilometres wide, protruding 200 metres out of the desert, with the Cooper Creek diverting around its base.

When the plane landed, Reg was told that Adria Downs Station about 150 kilometres to the east, just out of Birdsville, was reporting oil shows in Cretaceous marine sediments at 930 metres in their new bore. More great news. As soon as he returned to Adelaide he rushed to Bonython with the good tidings. But Bonython was preoccupied with business matters and the jubilation of the intellectual victory associated with the discovery did not stir his emotions like it did Reg's. His was a more prosaic interest centred on business prospects and commercial success. Whereas Reg was enthralled with the scientific process itself—the formulation of a hypothesis, the accumulation of facts, and the testing of geological theories in the hunt for oil.

A few months earlier Reg had begun writing an academic paper for the *Bulletin of the American Association of Petroleum Geologists*, in which he was outlining his theories concerning the petroleum prospects of the Great Artesian Basin. Now he had exciting new results to incorporate.

Bonython was more focused on the upcoming overseas trip planned for Reg and himself. Lev had interested Delhi, a mid-size independent oil company from Texas, in a possible partnership with Santos. Other companies including United Gas Corporation, Vacuum Oil and American Metals Climax Oil were also showing interest in joining forces with the small Australian explorer. Lev would shepherd Reg and Bonython to the appropriate offices and make introductions. The rest was up to them.

In September, October and November—while the first Santos well was drilled at Cordillo Downs—Reg played the corporate game in America: meeting oil chiefs over lunch; showing his photographs and reports; bartering farm-in terms, percentage interests and royalties; and feeling like

ROCK STAR

Santos Chairman John Langdon Bonython in Corpus Christie Texas, during negotiations with Delhi Taylor Oil Corporation, December 1958. The DC3, named 'Flying Ginny,' was the private plane of Delhi multi-millionaire owner Clint Murchison.

an exposed minnow in a sea of grinning sharks. Bonython had returned to Australia and the responsibility of the wheeling and dealing rested largely on his shoulders.

While some executives feigned disinterest, or pronounced Australia's oil market too limited, others could not help but be impressed when Reg displayed Jack's 1925 geological diagrams and his own photographs of the great Innamincka Dome. Consultant Frank Reeves, who had comprehensively analysed Australia's oil prospects on two trips to the country in 1934-36 and 1947-49 and pronounced them very poor, admitted that he had never specifically visited the central Great Artesian Basin and had only flown over Cordillo and Innamincka in a dust storm. He now lamented the obvious opportunities lost.

Reg returned to Adelaide at the beginning of December with a draft agreement with Delhi in hand and a delegation of four company representatives hard on his heels. Flying over the Cordillo, Haddon Downs, Innamincka and Betoota anticlines in a hired DC3 the Delhi men were awestruck and incredulous that others had been unaware of these great structures for so long. There was general high praise for Geosurveys and promises of lucrative contracts forthcoming.

By the time Reg had squired the reps around town on visits to the Premier, the Mines Minister and the Department of Mines, and packed them back on a plane for the US, he was exhausted. He had been working almost exclusively on Santos affairs for months without respite and was badly in need of a rest.

'Farewelled Delhi,' he wrote in his diary on the 18 December 1957, 'Xmas: took break.'

Chondroplon bilobatum

CHAPTER 7

ACROSS THE SIMPSON DESERT

THE NEW YEAR was consumed with Santos-Delhi negotiations. While there was a general feeling of optimism in oil circles for Santos's prospects, the reality was that money was tight. The oil stocks share-market boom was over and Santos funds were scarce. Bonython wanted to halt further drilling until the American men handed over their American dollars. Reg pressed for continuing exploration. Santos had the acreage and he believed it was important to retain the image of a confident Australian explorer. Despite presenting themselves as homely business buddies, Reg suspected some disingenuousness on the part of Delhi. He felt sure that Santos's potential Texan partners were smart and savvy corporate players. He was afraid Bonython would yield too much in his desire for cash in hand.

Frobilco, who had given up rights to the area only three years before, were now keen also to join with Santos in a renewed exploration effort. Chairman Maurie Mawby was deeply annoyed that his company had missed the enormous desert structures that literally broadcast the potential for oil underneath. But Santos now had all the playing chips. The game was to use them wisely.

Santos was not the only one feeling the pinch. Ampol's Bill Walkley was quoted in the paper as saying that 90% of oil exploration work in Australia would stop by the end of the year because of insufficient capital. The Rough Range-1 well that had heralded the oil boom had failed to produce in commercial quantities and after spending around £15 million on the search, the West Australian Petroleum consortium was steadily exhausting its supply of money and enthusiasm.

By mid-February, drilling at the Cordillo well—Santos's first in the Great Artesian Basin—had been stopped, much to Reg's dismay. But Delhi negotiations were proceeding. The company was talking up its Australian oil prospects, telling its American shareholders that Australia's crude oil market of some 175 000 barrels a day—all of which was currently imported—was an attractive incentive. The Santos area—slightly larger than Delhi's home state of Texas—was a good financial investment, they advised.

Finally, in May 1958, the agreement was announced. Santos and Delhi would carve up the Great Artesian Basin acreage in a chequer-board arrangement of alternating 940-acre squares. In this way, each company would have half of the total area, however, if one company found oil, the other could drill nearby on their own square with the hope of tapping the same reservoir. Santos would also receive a 5% royalty on production from Delhi's squares, and the government, 10%. Thus out of every 100 barrels of oil that Delhi might one day produce, 10 would belong to the government and five to Santos.

Delhi was also committed to drilling a deep exploratory well to at least 4000 metres, costing anything from $800 000 to $1.2 million, as well as spending $500 000 per year on exploration for four years thereafter. The agreement had been approved by both the South Australian and Queensland governments, as the Santos acreage, which now crossed state boundaries, was being treated as one huge prospect.

Bonython was delighted.

Delhi was now calling the shots and activity intensified. The company selected the Innamincka Dome as their first drilling target and subcontracted to the Department of Mines for a seismic survey and to Geosurveys for geological mapping of the area. When Reg produced the preliminary map that Geos's staff Heli Wopfner and Rudi Brunnschweiler had drawn there was astonishment at how much Geosurveys had already achieved. But Reg was noting some ominous signs. After all his dedicated work—from exploration through to contract negotiations—he was now feeling somewhat excluded from Santos affairs. Funding was still being cut, such that Geosurveys was actually extending credit to Santos and Geosurveys was relying on other sources of income to carry their Santos work. Reg vacillated between confidence in the future, and deep concern over Geosurveys's role in the new Santos-Delhi operation.

In addition to finessing the geological mapping, Geosurveys was also responsible for general logistics, including setting up a work camp about 40 kilometres north of Innamincka near Patchawara Bore, and blazing access tracks. Vehicle access to the area was completely random. The Strzelecki Track itself, from Lyndhurst to Innamincka was non-existent, save for an occasional grading of a passage through to Murnpeowie Station, 100 kilometres from its southern end. Mail and supplies for Innamincka Station itself were channelled through Broken Hill. Naturally Playford did not want New South Wales profiting from his South Australian oil initiative and he commanded the government Engineering and Water Supply Department, who administered roads, to get cracking on re-opening the Strzelecki.

The work camp at Innamincka would house Geosurveys, Delhi and Department of Mines geologists and geophysicists as they combed the Innamincka Dome and pin-pointed their first drilling target. Reg purchased a clutch of ex-public-transport double-decker trolley buses that his workmen gutted at the Woodville depot and transformed into sleeping quarters and offices and a fully equipped kitchen with upper-level dining for thirty. The

The cook at Delhi's Wadi work camp, north of Haddon Downs, stands in the doorway of Geosurveys's double-decker dining bus, September 1958.

yellow and silver buses emblazoned on the side with 'Geosurveys' would be towed to Innamincka in a convoy of 24 vehicles and caravans and 40 men, representing the official beginning of the Santos-Delhi oil search. The local press carried extensive coverage of the long crocodile slowly departing Adelaide for the expected five-day journey to Broken Hill, Tibooburra and finally Innamincka.

True to form, Playford, too, was now keen to visit Innamincka and see all the activity for himself. Reg decided to make a ground trip there, beforehand, to see the state of the roads and view camp operations. Griselda and the children came too. For six-year-old Margaret and three-year-old Douglas it would be their second long expedition into the Australian outback. They had already toured into the Gibson Desert via Ayers Rock,

ROCK STAR

Warburton, Mount Davies and Ernabella, as well as tagged along on trips to Wilkatana and Broken Hill. Doug was already a bushie through and through—more comfortable in a swag than a bed. When recovering from an eye operation while still a toddler, the attending nurse was perturbed to find that he refused to recover in the bed, preferring the surgery floor. Griselda reassured her that this was entirely normal. Each night at home Douglas was tucked into bed and later found snuggled under his eiderdown on the floor.

In early September the family departed for Innamincka following the dry bed of the Yandama Creek north-east of Lake Frome to Tibooburra, Reg checking for gas escaping from old water bores along the way. After a couple of days at the Innamincka camp site they carried on into the Queensland corner country to visit Heli Wopfner, who had been camped in the Haddon Corner area since July with his wife and two young children.

While a dust storm roared outside, the two families sheltered in Heli's Geosurveys caravan. The two women chatted while Reg and Heli traded work details and the youngsters played. Margaret was asleep in three-year-old Klaus Wopfner's bed. She was prone to various allergies and Griselda

Reg demonstrates gas ecaping from Coonana water bore to Margaret and Douglas, during a trip along the Yandama Creek to Tibooburra and then Innamincka, September 1958.

thought a rest would clear the red spots from her face. However, three days after the Spriggs left, Klaus developed the same red spots. When he developed a fever as well, Heli radioed the flying doctor and learned that not only did his son have measles, but so did people in Innamincka, Arrabury, Lake Pure, Planet Downs and Birdsville—the Sprigg family unwittingly leaving a trail of virus wherever they went. At Arrabury there were 12 people—the entire Debney family and the station's crew of stockmen—all sick at the same time, with the only person to care for them, old Elsie, who had survived the measles epidemic in the 1920s that had decimated her tribe.

'The thick end came when our little daughter contracted the measles,' says Heli. She was only one year old, with a 40-degree fever, a severe cough, short shallow breathing and hundreds of miles from medical help.

'Fortunately we had some penicillin so [the flying doctor] said, give her two-thirds of a penicillin injection, which my wife did. But of course giving an injection to a small child is something different. She punctured the poor child five or six times before she finally forced the needle in. The child was screaming and I was holding her and then my wife went out in the bush and just collapsed weeping on the ground.

'We kept watch all night and then at five o'clock in the morning she relaxed—the temperature dropped.' When the flying doctor nurse checked in at the specially arranged session one hour later, Heli was able to report that all was well.

'Those sorts of experiences—of cooperation, of help in the bush—this is something that you just never forget. These are the moments in a man's life which he is going to treasure,' Heli declares, almost 40 years later.

SOON AFTER REG arrived back in Adelaide he joined the Premier's party travelling up to Innamincka in a fully loaded DC3. The station could accommodate all the government officials and bureaucrats in their workmen's quarters but catering for them all was another matter. To solve the problem it was arranged for the South Australian Railways refreshment service to haul two caravans up there. The Engineering and Water Supply (E&WS) workers, used to making do in roadside camps, chuckled at the incongruous sight of cooks and waiters bustling about in such primitive conditions.

'Tom had the time of his life,' says E&WS regional engineer Laurie Steele. 'He was a great one for getting out in the outback—he thrived on that.'

The official party inspected the Santos-Delhi work camp and flew over the now-famous desert domes. In the evening Reg described the geology and showed his home-made films of the Santos operations. And once again Playford was pleased.

Preparations for drilling the first combined Santos-Delhi oil well continued. Another 50 men—geophysicists, drillers and engineers—would migrate to the Innamincka site in the new year, so that 100 men would be calling the remote camp home.

Workmen at Geosurveys's Woodville depot were frantically constructing and dispatching equipment to supply the expanded accommodations. However Geosurveys geologists were largely left idle, sidelined by Santos since Delhi had come on board. It riled Reg that his company was being treated this way; inconsequential; expendable; disposable. And yet at the same time, when Bonython was summoned to Dallas, Texas, in December 1958, to negotiate a farm-in agreement, who did he insist accompany him—Reg. His paper for the American Association of Petroleum Geologists had just been published to major interest from exploration experts. Bonython was just not up to describing the structural subtleties to US geologists clamouring for an explanation of this enticing new territory.

Their trip turned out to be a short but rancorous one full of arguments over royalties, retained interests, split-ups and farm-ins between the three members of the new consortium: Santos, Delhi and Frobilco. However, by Christmas the three were roughly one third partners in the Great Artesian Basin.

Meanwhile, the job of rehabilitating the Strzelecki Track got underway. Until the road was re-opened generators, refrigerators, a mobile powerhouse, drilling equipment…everything…had to be hauled the long way round, through Broken Hill. The Track had been left to deteriorate when the stock route closed in the 1930s and pastoral leases lapsed, but now the Premier's men attacked it with a vengeance.

It was undesirable to have a road-gang embark on such a devilish task in mid-summer, but Geosurveys was struggling; Delhi was complaining; the Premier was in a hurry: South Australian trucks must get through.

'It was most unfortunate we had to do the work in the middle of summer,' says E&WS engineer Steele. 'It was a job you normally wouldn't start on until the weather got cool because they're terrible conditions really, with the hot weather up there; trying to work.

'Anyhow when we had bashed a track through, a couple of convoys of equipment for the drilling of the bores went through. The first one was quite a small convoy and that was all right. But then they wanted to send a big convoy through. We'd briefed them on the conditions of the track and we'd emphasised it was a job for a four-wheel-drive vehicle or two-wheel-drives with assistance and lightly loaded. But they loaded up all this boring equipment—hopelessly overloaded, on most unsuitable vehicles, way over their axle-loads of eight tons. And of course, as soon as they got in the Cobblers they all got stuck,' he laughs. 'They weren't properly equipped. They had no gear for pulling themselves out of the sand, not enough food, not enough water, and things were pretty ordinary.

'In the end we were instructed to give them all the assistance possible. We had our dozers and graders and what have you, and we had to pretty-well pull every vehicle through the Cobblers for about 20-odd miles—just hook them up and just drag them through the sand.

'It would've been about three months, from go to woe, to put the Track through. Originally we were only supposed to build a track for four-wheel drive vehicles but I think the thing just snowballed. We had to keep on improving it and claying the worst spots with decent clay, because unfortunately in the whole length from Monte Collina to Innamincka—about 150 miles—there's not a stone or gravel or anything at all. It's just sand and clay. So all you could do when making a road—the dozers graded; and then the worst portion, spread clay over it. If you had any water of course you could water it in, but most times you just clayed it and mixed it in with the sand and you could get a reasonable road out of it.'

The Premier didn't have to suffer the privations of the road trip when he flew to Innamincka in late February 1959, as guest of honour at the Santos-Delhi 'breaking ground' ceremony. On 28 February, on the eve of Reg's fortieth birthday, a couple of dozen dignitaries, reporters and photographers scuffed up their shoes in the powdery, red dust to see Playford drive in the peg to mark the Innamincka-1 well. When he had finished he presented the hammer to Reg.

'We trained Reg but he left us,' the Premier had announced earlier, putting his arm around Reg's shoulders. 'Some day we hope to get him back.'

Like the first signs of oil at Wilkatana, the day at Innamincka remained in Bonython's mind as one of his favourite memories.

'It was a terribly hot day,' he recalls 25 years later. 'We got back to the plane about dusk. Because it had been standing in the sun all day, the plane was extremely hot. We were hot. Nothing was available to drink. The pilot said, "Get in. When we are up in the air, I'll radio to Leigh Creek for liquid refreshments."

'An hour or so later we arrived at Leigh Creek, in the dark. We got out of the plane to stretch our legs, to enjoy the relative coolness and to watch with gratification as a couple of hand trolleys approached the plane and loaded various forms of liquid refreshment. The pilot said, "Get in. We'll take off and enjoy a drink on the way to Adelaide." The plane got air-borne, and it was then discovered that all the beautiful liquid refreshment had been loaded into the cargo area—and there was no communicating door.'

WHILE REG RETURNED to Adelaide with the official party, Heli Wopfner stayed behind. He would remain on site as the official Santos well-site geologist. Delhi and Frobilco would also have their own men on the ground to keep tabs on progress. A couple of weeks later the rig was raised. It was the very same rig that had drilled the Rough Range-1 well. The 40 metre-high-tower that had once stood on the shores of Exmouth Gulf, now rose into the sky from one of the driest spots on earth. On the 29 March it was 'spudded' and drilling of the deepest well in Australia began.

Within a week the well was down to 563 metres. A month later there were the first signs of hope. Traces of wet gas at 1200 metres and then an 'oil show' at 1232 metres, with the well just over a quarter of the way to its target depth. But a few more such showings was all there ever was. At 3854 metres drilling was halted and the hole was finished off as an artesian water bore. There was no oil gusher under the great Innamincka Dome after all.

It was disappointing for geologists around the country who had watched progress with interest. But Reg was not disheartened. To him the results spoke volumes: that there were sedimentary beds of the right age under the Great Artesian Basin, and furthermore they were oil and gas prone. In his opinion they had learned a lot from their first test. It was far too premature

South Australian Premier Tom Playford (at right of sign) holds the hammer used in the 'breaking ground' ceremony for the Innamincka-1 well, 28 February 1959. The hammer was presented to Reg and now hangs in the dining room at Arkaroola.

to disparage the whole area. This was a wildcat well, after all, and experience in America had shown that the chance of success with wildcats—any well put down in a completely new, untested area—was only 10%. To discover a small commercial oil field of a million barrels a company may have to drill 50 wildcat wells. To discover a field of 50 million barrels, up to 3000 wildcats would have to be drilled.

Delhi executives, too, were experienced enough to know this. Besides, they had committed to drill at least one well to 4000 metres. They shifted camp to the Betoota-1 well site 200 kilometres north and raised their rig again. But after four months, that too was declared a dry hole. It was time to regroup.

Reg, too, was considering his position. Santos would keep drilling and oil would one day be found, he was sure of it, but his role would be one of a minor shareholder cheering from the sidelines. His pioneering part in Santos was over. Like Radium Hill, he had defined the deposit and nurtured its development, only to have the reins snatched from his grasp when bigger interests muscled in. The rejection hurt, but maybe it was somewhat of his own making too. He was no white-collar 'Grenfell Street miner.' He knew that. He was not suited to life in an office where the closest you got to faults and anticlines was a map on the wall. He was still happiest with his feet on the ground—the real ground; rocks and sand, slates and shales. He was not ready to retire his khaki shorts and geological hammer just yet. For him the corporate side of business—the stock market playing and political preening—was there to fund the fieldwork and exploration, not the other way around.

So he went stumping around the traps and found new clients to fill the Santos void. The LH Smart Oil Exploration company held rights over most of the area east of the Santos lease so Geosurveys contract teams were sent to Quilpie and Eromanga near the Grey Range of south-west Queensland on its behalf. Geos staff were also hired out to companies operating around Tamworth in north-eastern New South Wales; north of Longreach on the plains of Central Queensland; near Mootwingee in western New South Wales; the Gulf of Carpentaria's Mornington Island; and in South Australia's south-east region—an area which had long held interest for Reg. Geos staff were also on the ground at Mount Davies where the nickel operation was in full swing.

The International Nickel Company had delivered a favourable assessment of the Mount Davies territory in north-western South Australia. Their experts had advised that the nickel in the area, extending into Western Australia's Blackstone Range, was not a sulphide deposit, but a low-grade oxidised form of the metal that was more expensive to purify. However, the area was vast and there was potential for sulphide nickel, chromium and other platinoid minerals yet to be found. Nickel, itself, was highly valued in the munitions industry for the production of hardened armour plate. There was also a rapidly expanding market for stainless steel production and industrial applications for alloys and nickel-plating. The world's largest nickel producer could not afford to let such a lucrative deposit fall into competitor hands.

The Canadian company had offered to combine forces with Reg's company, Nickel Mines of Australia, to form a private joint venture: Southwestern Mining Company. International Nickel was paying for all the exploration and initial development. Because of the extreme remoteness, costs were high. Geosurveys, naturally, was conducting the fieldwork.

Reg's pioneering organisation, which he had launched seven years ago with his mortgaged home and two employees, was now the largest geological and geophysical consulting and contracting organisation in the Southern Hemisphere. Geosurveys operations in his Pop's backyard at Woodville had developed into a full suite of administrative offices plus workrooms and technical offices for Naturelle Jewellery and Specimen Minerals, and a full mechanical workshop and storage area. He had more than a hundred staff on his payroll and felt the relentless responsibility of keeping them all employed. While managing the various geological contracts consumed most of his time, Reg's brain was never idle and he was always dreaming up new business schemes and new ways to keep his staff occupied.

He launched a quarterly magazine called *The Australian Amateur Mineralogist* and a monthly newsletter called the *Australian (Petroleum)*

Scouting Service that summarised drilling and geophysical operations across Australia and New Zealand. His company Naturelle Jewellery, was the first Australian endeavour to sell polished gemstone jewellery. When his prospectors returned with bags of jasper, amethyst, rhodonite and beryl, Geos staff at the Woodville depot poured the stones into tumbling machines and polished them to a high gloss, following written instructions from Reg's US advisor. Silver mounts were attached to form pendants, bracelets, brooches and earrings that were marketed as 'uniquely Australian' accessories.

'Australians just don't realise that these gems exist in their country and people often ask me if I bought this bracelet on the continent,' Geosurveys/Specimen Minerals employee Mrs Roma McGlinchey told a reporter from *The Advertiser*. 'The articles are much in demand as souvenirs and as gifts for people overseas. We are not aiming to replace mulga wood objects as a tourist attraction but we do think we are giving people an alternative.'

Other mineral and fossil specimens were compiled into kits and sold to schools and universities through a side-business called Specimen Minerals that specialised in educational aids.

But finding oil was still the main game. When one company defaulted on payments, bequeathing its acreage to Geosurveys instead, Reg decided to go all the way and apply for an exploration permit of his own. Now his seismic teams had the north-western Simpson Desert to explore for themselves. But being lease-holder and contractor for the same plot of land could become messy. Perhaps he should form his own exploration company: A publicly funded wildcatter that could search where he thought best. He would begin with the coastline, he decided. He still believed there was oil down in the state's South East, and the continental shelf of Adelaide's Gulf St Vincent was completely untouched. He would call this new company: Beach.

BEACH PETROLEUM WAS incorporated in December 1961. Directors of the company were Heinerich Carl Meyer; John Matthew Dwyer; Lancelot Raymond Ackland; Geosurveys's secretary and accountant, and former Radium Hill clerk, Rennie Francis Middleton; and Reg's schoolfriend now

ACROSS THE SIMPSON DESERT

lawyer, Ernest William Palmer, who was Chairman. Reg was Geological Consultant to the new company and Geosurveys, of which he was still Chairman and Managing Director, was appointed technical manager. The company commenced active operations in February 1962 with a two-pronged approach: probing the South Australian coastline, and unravelling the mysteries of the Simpson Desert.

Investigations in the Simpson would begin with a detailed gravity survey over the entire lease area of almost 6 million acres. This would be a complicated operation that required careful planning. Four parties would embark into the Simpson: three entering from points north and travelling south to meet up along the way with the main party crossing from east to west. Eleven Geosurveys personnel, under contract to Beach, would comprise the expedition. Reg would lead the main party, his crew comprised of Mrs Griselda Sprigg, Miss Margaret Sprigg and Master Douglas Sprigg.

The press gasped at the idea of Reg taking his young family into a sandy wasteland of 100 000 square kilometres. But those who knew Reg took it in their stride. 'That was Reg,' says Betty Wilson. 'That was the Spriggs.'

Darby von Sanden, Geosurveys's recently employed Operations Manager, would lead the 'S' team consisting of he and his 18-year-old son Anthony, and two other pairs would complete the 'J' and 'Q' teams. When the Spriggs set out from Andado Station driving eastward into the desert, the three other teams would already be on the ground, steadily working their way southwards to an imaginary horizontal line through the centre of the desert. They would each be taking gravity measurements along the way to construct an overall grid picture of any geological anomalies buried beneath the shifting waves of sand. The gravity meter carried by each team was a glorified weight attached to a spring. At each survey station, the operator would place his gravimeter on the ground and measure any extensions or compressions of the spring corresponding to minute changes in gravity. The gravity reading changed according to changes in density of the underground structure. Thus a gravity change could indicate the presence of subterranean folds and domes that had the potential to trap oil, or just as importantly,

153

an unprospective intrusion of granite or a thin veneer of unpromising sediments cloaking shallow basement rock.

Taking gravity readings was time-consuming and tedious, but routine work. For each team their most challenging task would be to rendezvous at the correct time and place with Reg and his family in their Nissan Patrol. Four small vehicles roaming alone in the desert, hoping to meet each other in an endless sea of sand. There would be air support but it was a daunting task. A mechanical breakdown could spell doom; an error of navigation, fatal.

In typical Sprigg style, Reg was excited. This was true exploration. He would be going where only two parties had travelled in the history of the white man.

In May 1936, Ted Colson a grazier from Blood Creek Station on the western fringe of the desert, and his Aboriginal companion, Peter, had ridden across on camels in just 12 days. Colson had purposely waited for a favourable wet season and his account of his journey, for the *Queensland Geographical Journal*, describes an easy jaunt through dense masses of herbage all the way. His notes convey none of the spectres of isolation and thirst that foiled other attempts.

'After going on to Birdsville, where I spent a very pleasant two days I returned over my own tracks west…Thus ended a pleasant experience which I hope to repeat next year. I will then have crossed and taken stock of 12 000 square miles of what was a blank space on our maps, and to do that is all the reward I seek.'

Three years later, in June 1939, Reg's former university professor Cecil Madigan and his nine-man team crossed with camels in 32 days conducting a scientific survey of Australia's 'Dead Heart.' Now, Reginald Claude Sprigg would do it in a motor vehicle—a legitimate first for the post-World-War-Two era of modern exploration.

On Thursday the 30 August 1962, Reg packed his family onto *The Ghan* at eight o'clock in the morning for the start of their desert adventure. He would fly up to Andado Station, 70 kilometres east of Finke just over the North Territory border, to meet them the following afternoon.

At the station homestead on the Saturday morning Griselda packed. Swags; a medical kit; two wooden tucker boxes; a crate of oranges and apples; blanket-filled draw-string bags; two heavy, metal ammunition trunks containing canned food and cooking utensils; and several cardboard cartons, were squashed into the bulging short-wheel-base Nissan. She crammed all this around the 44-gallon drum of petrol and two 20-litre water containers—the family's six-day supply—that consumed the vehicle's rear storage space. Reg stood back and marvelled. She seemed to have included everything except the proverbial kitchen sink; everything that is, except his camel-grip containing field clothes, photographic film and such likes, which seemed to have gone missing en route from the railway siding at Finke.

At two o'clock in the afternoon Reg and Griselda slid onto the front seat; the two children between them—four Spriggs shoulder to shoulder with spirits high. They set off south, heading for the South Australia-Northern Territory border running along the 26th line of latitude.

On station property—Andado and Mount Daer—there were tracks to follow south and east, but by day two they had left the tracks behind—fat cattle replaced by swift-footed ground lizards and flocks of green shell parrots; the compass their only guide.

For Griselda, Margaret and Douglas it was their first experience of sandhills and spinifex mounds—giant versions of both. When Reg charged up the western slope of the first red dune, Griselda implored him to stop at the top. The steep eastern dune face caved away alarmingly and she was afraid to ride in the Nissan to the bottom. While Reg captained the vehicle—careening down at crazy, swooping angles to the flats below—Griselda and the children scrambled along behind, feeling sure the car would soon turn over. Reg was most obliging, as one formidable-looking sand-dune rose behind another, but Griselda knew that she would have to overcome her fear.

'Reg is very tolerant about this, but I feel sure his tolerance will not last more than the first day,' she wrote in her diary. 'I knew it would be rough going, and Reg warned us in advance that this would be no picnic.'

The spinifex was almost as bad as the sand. Great wide hillocks of the stuff that entangled the wheels of their vehicle and sometimes beached them on top, leaving them stranded with a scorching exhaust pipe threatening to set the spiky mass alight. Griselda and Reg would fly out of the car, frantically scraping with the shovel and grabbing handfuls of spines with their bare hands to clear the grass before the smouldering took hold. The thought of the 44-gallon drum of petrol sitting above them giving extra frenzy to their efforts.

At some of these unscheduled stops, Reg assembled his movie camera and filmed buttercup-yellow daisies and yellow and white billy buttons waving in the fine orange sand on the dune tops while Margaret and Douglas frolicked like exuberant puppies. Griselda even joined in when they slid down the steep soft faces on imaginary toboggans, scooting along on their bottoms, leaving deep orange furrows behind them.

At midday Reg called a 20-minute rest for lunch. By mid-afternoon they all needed another break from the incessant bumping and bouncing. Backs and bottoms were bruised; lips and skin parched from the hot north wind; sweat smeared their faces and trickled down their torsos. Griselda took action. First she slathered everyone in thick moisturising cream, then stepped out of her slacks and blouse.

'My it's better now, sitting in my nylon panties,' she thought.

The kids quickly follow suit; driving along in your underwear—what fun! Only Reg held out—persisting with his corduroy slacks and thick shirt, since his desert clothes were in the lost bag.

'He makes me sweat to look at him,' Griselda confided to her diary. But after digging the vehicle out of one more sand bog, he too had reached his limit. He looked quite a sight taking a compass reading a short time later in singlet, boots and underpants rapidly turning from white to red with desert rouge.

By the end of the first full day of Sprigg Simpson Safari they had crossed 108 sandhills in 35-degree heat at a maximum travelling speed of seven kilometres per hour. It was a unique way for Reg to spend Fathers' Day. They camped that night on the crest of a high red dune, with the desert

ACROSS THE SIMPSON DESERT

Reg had to strip to his underwear for comfort when his field clothes went astray at the beginning of the family's safari across the Simpson Desert, September 1962.

panorama at their feet and a sparkling view of Venus in the western sky framed by a thin crescent moon.

Before last light, Griselda ruminated on the day they'd had while Reg and the children took their revenge with spinifex bonfires in the valley below.

'Well this is indeed as near to heaven as I will probably ever get,' she wrote, 'and it is pretty jolly good. My back is rested. I am fed, and now with a cup of coffee and the long awaited cigarette, I wouldn't call the King my Uncle. What unfortunate people the city dwellers are who know so little this bliss. Few are privileged to enjoy the peace and quiet of nature at its best here in a good season in the desert. At least my children will know Australia before they know other parts of the globe much better. In this vast continent there is much beauty. Why go to the other ends of the earth first to find it?'

THE NEXT MORNING the family rose at sunrise, as usual on these outback trips. There was a buzz of anticipation in the air for today they expected to rendezvous with the first of the Geosurveys teams travelling south from their original campsite at the Hale River. By eight o'clock the breakfast dishes were done—digging them into the sand and polishing them clean—and they were on their way.

There were fewer sand bogs that day and more time to enjoy the scenery. The country had received unexpected good rains in early winter and the desert was a profusion of wildflowers—'a botanist's delight,' Reg called it. He was as captivated by the waving masses of yellow and white everlasting daisies interspersed with purple parakeelya and yellow munyeroo that carpeted the inter-dune corridors, as he was by the unique dune formations—with their steep easterly faces and more gentle western slopes—and what that told him about prevailing winds in the last glacial ice age. Budgerigars and zebra finches flitted about while he took soil samples with his auger. Grey, wren-like birds with red under the wings used the 'broken wing' ruse to divert attention from their silky nests with three speckled greenish eggs inside. All this he absorbed and noted with avid interest. Nature was truly fascinating, he marvelled.

The children sang the day away and took turns clicking off the sandhills on the dashboard-mounted pedometer. As the shadows began to lengthen, however, and they still had not converged with the 'J' team, morale began to dip. And then there was a shout. Griselda, eyes straining behind her spectacles, had spotted familiar tyre tracks crossing their path. Their boys had obviously passed by not long before. The first sign of company—what joy! The mere thought of their Geosurveys colleagues being somewhere nearby made the Sprigg party feel a little less lonesome in the midst of such emptiness.

They made camp and Reg positioned the car with its headlights pointing south. After dinner he flashed the lights repeatedly and then, on the dot of eight o'clock, fired the first signal rocket. Margaret and Douglas cheered as the ball of yellow light raced into the night sky with an impressive 'whoosh' and exploded overhead. Within a minute there was an answering rocket about eight miles to the south-east. They had neighbours!

By next morning the 'J' team had backtracked to their campsite and the first rendezvous was complete. Before they arrived Griselda had combed her hair for the first time in three days, sacrificed some precious water to wash hands and faces all round, and applied some lipstick.

'Better not to frighten the boys too much at the start,' she thought. 'I wear no makeup whatsoever on these trips, though I would never be seen dead without lipstick at home.'

Despite not routinely wearing makeup while camping out, she nevertheless took some along, like emergency rations, and to date, lipstick and eyebrow pencil had both come in handy to label road junctions at gravity stations. On a previous trip her talcum powder had been used to lubricate a tyre rubbing against a makeshift axle. Sometimes it paid to have a woman on board.

They set off again—a convoy of two now, the Spriggs in the Nissan, and Geoff Rowley and Gordon 'Fred' de Rose in the International Scout. Freddie and his squat, two-door utility were officially on loan from the International Harvester Company of Australia. Before departing on the expedition Reg had approached agents for Toyota, Nissan and Scout four-wheel drives, in the hope they would provide a free vehicle to test their products' reliability, ability and endurance. International Harvester had sent along Freddie—service station specialist and experienced bushman—to equip their vehicle with its best chance of success. Already he had had to replace a broken axle with one of the many spares he had brought with him.

'This blighter is far too light for this sort of country,' he confided to Griselda, as he slapped the side of the Scout. 'I had a feeling it might be.'

He didn't draw attention to the fact that the Scout was seriously overloaded with full drums of fuel and water, in addition to crates of tools, spare springs and axles. But if anyone could shepherd the Scout through to Birdsville, it would be Fred.

Meanwhile he and Geoff had also brought news that Darby, too was having vehicle troubles. The Geosurveys plane had deposited Darby and

Anthony on an isolated claypan further east to take charge of an old Toyota LandCruiser nicknamed, 'Yellow Peril' that had been taken up there months before. It was an early version of the LandCruiser with no transfer case and it already had four broken shock absorbers—an unlikely vehicle in which to embark on a solo expedition into the desert. Now it seemed that the Yellow Peril had 'lost' both first and reverse gear.

'The corridors are so choked with spinifex we've been getting jacked up every 100 yards or so,' Darby reported when they reached him on the radio. Everyone at the Sprigg end nodded in recognition. 'To make any progress at all we've been driving along the dune crests. We were climbing the face of one of the bloody hills when we lost first gear—sheared the rivets joining the gear to the drive shaft. Reverse gear's connected to the same drive, so we've lost that, too.

'Don't worry,' he continued, sensing the concern amongst the group gathered around the radio, 'Whenever the corridors get too choked, we just charge the dune faces flat out in second gear. The rate we're going we'll make the rendezvous point about a week before you get there!'

It was such a malfunction that Reg was surprised to hear Darby saying that he intended pushing on nevertheless to meet them at their most easterly rendezvous point and then carry on together to Birdsville. Perhaps the wiser option would be to return to the claypan landing strip and wait for a pick-up. Darby was resilient and resourceful but still, Reg doubted he could mollycoddle the Yellow Peril along for the next 80 kilometres. However, there was nothing Reg could do to assist him from here. His job was to make sure his own convoy made it to the next rendezvous point on time.

He studied his survey positioning charts and compass readings and when, during the days' travels, he encountered topographic features like watercourses, claypans or salt lakes that he recognised from his aerial photographs he marked these on his map. He was also taking gravity readings each day, like the other team members, to tie the survey data together. By the end of the day his arms ached from wrestling with the steering wheel and fighting to keep control as the Nissan bucked and twisted over spinifex clumps and slithered down sand slopes, hitting the hard-packed floor of the interdune corridors with a bone-jarring thump. Around the salt lakes the dunes were often hillocky, with desert grasses adorning sandy moguls like those formed on a steep, bumpy ski run. It made for a demonic ride with the steering wheel always threatening to jump out of your grasp. Sometimes the wheels would spin and fountains of sand would shoot up the sides of the vehicle. And then the tyres would find traction and the vehicle would lurch forward once again. Away from the salt lakes the dunes were powdery smooth slopes down which the vehicles skidded slowly with a bow wave of rusty sand at the front.

They were long, exhausting days but so far Reg was pleased with progress. 'A goodly place to camp for the night,' he noted in his diary on Tuesday 4 September. 'So far the survey is going well.'

By day five they had travelled 150 kilometres from the Finke river 'turn off.' They were one third of the way to Birdsville and right on course. That night they met up with the 'Q' team of John Platt and David Hughes also travelling south-eastwards from the Hale River base camp. The Geosurveys Cessna, too, was sitting on a large claypan at Platt and Hughes's campsite. Four of the five desert teams were together. Furthermore pilot Colin Semmler had Reg's long-lost camel-grip with him, which had jumped out of the Landrover on the way from the train at Finke to Andado station. That Friday, the 31 August—less than a week ago—now seemed so long past.

It was a happy reunion for Reg as he dug out a pair of shorts, spare 16-millimetre films and a fresh set of underwear. Griselda decided it was about time they all had a change. Due to the continuous rubbing, their smalls all had 'perforated rudders,' so they burned them gleefully on a ceremonial funeral pyre. Reg's clothing comfort was short-lived, however. Now one of his boots was missing. Teetering about in one of Griselda's Italian sandals would have to suffice. Griselda shook her head at the sight of him, and Margaret and Douglas giggled. After tea, Colin broke the desert silence with his accordion and the nine explorers serenaded the camp fire to celebrate their togetherness.

It was rough travelling on day six but Reg would not let up. He was anxious to make the evening's rendezvous with the von Sandens, or to organise a rescue mission if they failed to appear. He urged the Nissan over the unrelenting spinifex clumps like a man possessed. Dune after dune rose up to block their path. The peaceful acceptance of days past seemed to be replaced with a skein of frustration that had enveloped the car. Griselda was attempting to shelter behind a tea towel rigged up as a curtain to shield her from the sun. The day before, her left hand had become badly sunburned holding the door closed and it was now painful and swollen. The children too were uncharacteristically grizzly. Finally, at one o'clock in the afternoon, Griselda persuaded Reg to stop. It was eight hours since they had last eaten and hunger could account for much of their malaise. Young Douglas, in particular, was known to flag when his stomach was empty.

After a quick break they pushed on. Tensions again mounted as the light began to fade and still no sign of the 'S' team. And then, there it was: The Yellow Peril in the gloom, parked boastfully in silhouette on top of a 25- to 30-metre-high sand dune—one of the biggest about. Reg flashed his lights and an answering burst of fire sparked on the hill.

'What a thrill!' Griselda thought—'Strange how such little things can take on so much meaning in this environment.'

They could see them clearly now; Anthony waving on top; Darby racing down to meet them, long arms and legs flailing like a giant praying mantis. They marvelled at how Darby had managed to climb the massive dune in his crippled vehicle, and bundled out of the Nissan in a joyous ball of relief. Backslapping and handshakes all round, and good-natured teasing such as: 'What took you so long?' and 'What kind of a camp site do you call this, anyway? Bloody freezing; no wood…' It was wonderful to be together.

When the Scout and the 'Q' team's green Toyota LandCruiser trailed in to join them the Scotch was brought out to toast their success and stave off the desert chill.

'This was a moment of real pleasure,' Reg wrote in his diary. 'Pleased we are indeed for the safe rendezvous of all parties absolutely on schedule. I personally feel very relieved that all cars have made the distance so well and rendezvoused right on the hour. It could have been quite different.'

THEY WERE NOW halfway through the Simpson Desert but still there was a long way to go.

Next morning the camp was bustling. Freddie crawled under each of the vehicles tightening bolts; Reg—now in a pair of Darby's old desert boots—issued radio instructions to the Geosurveys pilot; a Geos trig sign was raised on top of the sandhill; gravity meters and barometers were checked; and cars refuelled. Griselda packed up camp. They set off in a convoy of four for the first time, and by the end of the day had put another 42 kilometres and more than 170 sand dunes behind them.

They camped that night at the edge of a big, dry swamp—a welcoming wide flat with generous stands of gidgea inhabited by twittering zebra finches tending stick nests that each housed four small white eggs. Reg figured that their camp was about 50 kilometres from Poeppel Corner—the intersection of South Australia, the Northern Territory and Queensland. He was keen to locate the official corner peg and erect a Geosurveys trig station at the site. It was 82 years since the surveyor Augustus Poeppel had completed his epic survey, laying his measuring chains across the middle of the continent to mark the South Australian border. It would be an honour to stand in his footsteps.

Two days, another 300 sand dunes and five salt lakes later, they spotted aircraft wheel marks and a large cross scraped into the ground. In the centre of the cross was a wooden post with the faint numbers '186' burned into its surface. It must be Poeppel's post, the numbers corresponding to the mileage west from Haddon Corner—the most north-easterly point of South Australia.

The wheel tracks belonged to Graham Wright's Piper aircraft. Graham was a Geosurveys surveyor and pilot who, with his intrepid wife, Trish, crisscrossed the desert gathering the data with which to construct Reg's critical photomosaic maps. It was a life lived by the stars and Trish loved it. They would study the star almanac, make a primitive camp and then sit back and wait for the stars to appear. While Graham pin pointed the stars' locations with his theodolite, Trish recorded the details of latitude and longitude, bending over her notebook in the dimmest torchlight.

'It was hot as hell during the day, and cold at night,' she remembers. 'You couldn't have a fire because of the refraction and you couldn't have any light at all. Sometimes you didn't finish before three o'clock in the morning. Gosh it was cold. But I loved it.

'It was very satisfying because we did it from start to finish. We'd do astro-fixing for a few months—get a fix, then go a few more miles and do another fix. Then we'd get a plane, take the door off, find our crosses on the ground and photograph from three different altitudes. The lowest altitude was very low—always at the stall-point of the plane. I was often holding Graham's feet and flying the aircraft because the camera was very heavy. You had to hold it with two hands—one on each side of the camera, and the lens had to be directly above. Sudden down-drafts were a danger. But we never had an incident.'

Some of the astrofix locations Graham and Trish could drive to in their yellow Geosurveys Landrover. Other spots could only be reached by air. Like Colin Semmler and other experienced bush pilots, Graham would find a likely looking claypan or salt lake and test the surface by touching down cautiously. Any stickiness or sucking sensation on the wheels and he'd immediately power away, hoping to escape before becoming bogged. It was risky flying, but they all did it. There was no other way.

At the site of Poeppel's 186-mile marker post Reg and the Geosurveys team now set to work building an impressive trig station. The boys stood a 6-metre galvanised pole in an empty 44-gallon petrol drum and filled it with sand, then stabilised the contraption with guy ropes fastened to angle iron posts embedded a couple of metres into the sand. At the top of the mast they rigged a vitreous enamel Geosurveys circle. This marker was meant to last.

After an hour or so of effort, while the finishing touches were being made, Reg took off on a scouting trip in the Nissan. Driving along the very edge of the western shore of Lake Poeppel he spied a thick, wooden post lying flat on the ground. It was rotten and termite-eaten but it was undoubtedly Poeppel's corner peg. Graham Wright's astrofix and the Geosurveys's trig station were in the wrong place—a third of a mile to the north-east. This spot where he was now standing, was the true intersection of South

Margaret and Douglas pose with the original Poeppel corner post marking the intersection of South Australia, Queensland and the Northern Territory, discovered 9 September 1962, during the first vehicle crossing of the Simpson Desert.

Australia, Queensland and the Northern Territory. Poeppel's 82-year-old triangular post—just over a metre long, with the state names emblazoned on each face—was the proof. But the post was in poor condition. Most of the wood in the centre was a decayed honeycomb and Reg felt that it would soon crumble away completely. He thought it could be treated and preserved in Adelaide, so decided to take it with him. The relocated Geosurveys trig would mark the site in the meantime. Thus Reg Sprigg was the last person to see Poeppel's peg in situ.

The journey east was now assisted by the occasional ancient mile marker post to follow. Some stood tall still—marching eastward along the 26th line of latitude; others, like the corner peg, lay down in submission to the elements.

The expedition was nearing its destination. Reg was sure that the 140 kilometres from Poeppel Corner to Birdsville could be accomplished in one—long—day if necessary, providing all went well. But the dune bashing had taken its toll on the vehicles, the little Scout most of all, and at 4.30 in the afternoon on day 10, there was a loud bang.

The convoy came to a halt and Freddie slithered under the Scout to find that the rear differential was broken—his four-wheel-drive vehicle was now a two-wheel drive. Disheartened but not defeated, he and Darby removed the tail shaft, transferred most of the load to the other vehicles, and pushed on—literally. The crippled car successfully negotiated the next two sandhills but only conquered the third with David, Geoff and Griselda shoving from behind. And then the front axle broke. The Scout had finally surrendered.

Even Freddie realised she would have to be abandoned this time. The new car that the International Harvester Company hoped to promote as a hardy four-wheel-drive, would not even make it home. They were now so near to Birdsville—had come so far—it was painful for a bush mechanic such as Fred to leave her behind.

It was a dejected party at camp that night.

'Poor Fred is really depressed,' Reg wrote in his diary. 'It is he more than any other who has kept all the vehicles going and that it should be his vehicle that packed it in first, is fate at its unkindest. Actually this lower-powered vehicle has done a remarkable job and had a really heavy load.'

'Doesn't seem right,' Freddy said, 'The little bugger giving up just as the hardest part is nearly done.'

Everyone felt bad...and that set them all to thinking. It was only about 120 kilometres to Birdsville and they now had graded tracks ahead of them—trails from the east forged by other oil exploration teams, a maintenance track along the old netting fence along the South Australia/Queensland border and soon there would be station tracks on Alton Downs to follow. Reg decided it was worth trying to 'tow' the Scout to Birdsville by replacing the Scout's front axle and linking it behind the good Toyota

ACROSS THE SIMPSON DESERT

Reg and Darby provide a back rest for each other at a campsite in the Simpson during the first vehicle crossing of the desert in 1962.

with tyres lashed between the two vehicles—a kind of six-wheel drive road-train.

Darby, Freddy and John Platt set off into the darkness in the green Toyota to instigate the plan, and re-emerged next morning, triumphant.

'Next thing we saw the Toyota and the Scout linked in tandem coming madly down the nearby sandhill,' Reg wrote. 'Darby had made it with the wired-tyre technique. Darby and Fred were truly flushed with success. It was a good effort.'

The party then decamped for the final time and headed east for Birdsville. The luxury of track travel was blissful after 11 days of bush bashing; a wash at the Alton Downs bore drain, sublime.

'The water is hard and the soap will not lather, but the feeling of water on my body for the first time in 11 days is a heavenly experience,' Griselda wrote.

Another 110 kilometres and they reached Birdsville at 9.30 pm on Tuesday, 11 September 1962. Eleven days; 280 miles; 1386 sand dunes: the

Simpson had been crossed for the first time by car. It was not the jubilant entrance they had hoped for—due to miscommunication the party had become separated at the last moment, with the expedition now traipsing into Birdsville in dribs and drabs—Griselda, the children, Fred and John in the Scout and the green Toyota in the evening; Darby and Reg in the Nissan at midnight; and Anthony, Geoff Rowley and David Hughes in the Yellow Peril the next morning. But still it was done.

THE NEXT DAY there was a congratulatory telegram from Mr Alexander in Adelaide, President of the Royal Geographic Society of Australasia. Reg also paid tribute to the source of his inspiration—his former Adelaide University lecturer, Dr Cecil Madigan—by building a cairn at a corner of the airstrip opposite the Birdsville Hotel, to commemorate Madigan's 1939 expedition. Madigan's son, Russell had flown in from Melbourne to participate in the cairn building and speeches, and the gala dinner that evening. For the festivities, Griselda donned her beloved makeup and a gleaming white Grecian dress trimmed with gold thread, which she had transported all the way across the desert, sealed in a plastic bag.

The following day Reg was in the Cessna and home in Adelaide a few hours later, leaving his family to complete an expedition addendum by driving home 1220 kilometres down the Birdsville Track to Adelaide with the remainder of the Geosurveys team. Reg had business to attend to. Beach's Mount Salt number one well in South Australia's South East was at a critical depth, he had a report to write on the Simpson survey for Beach, and all of his other projects needed his attention. This was no time to bask in his achievement and solicit plaudits. There was work to be done.

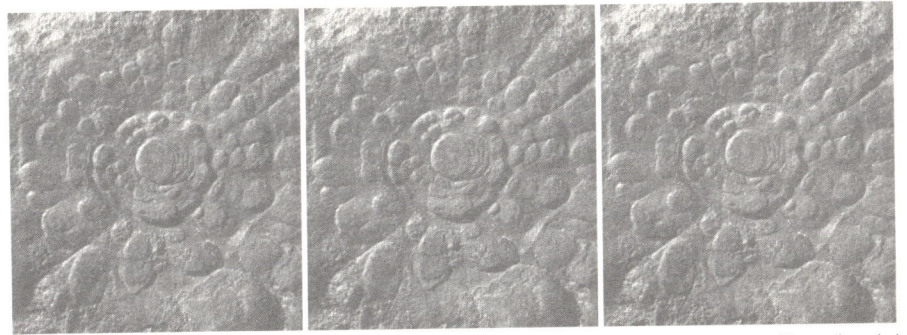

Mawsonites spriggi

CHAPTER 8

WATERWORLDS

GEOSURVEYS'S SURVEY OF the Simpson Desert had provided Beach with gravity information over a 20 000-square-kilometre area, averaging 160 kilometres from east to west and 130 kilometres from north to south. The gravity teams had made traverses along interdune corridors 8 kilometres apart, taking measurements at every one and a half kilometres within each corridor. It was a mammoth effort in completely virgin territory. A gravity operator ferried around in a helicopter would take the final few measurements to complete the entire grid and the data would then be interpreted by Geosurveys's Chief Geophysicist Dr Wilfred Stackler and his support team of female 'milligals'—a term of endearment based on the official gravity unit of measure, the milligal.

While Reg and his crews had been occupied in South Australia's dry interior, other Geosurveys teams had been busy along the coast. His geologists had conducted a series of seismic tests along the foreshore at Glenelg, Adelaide's premier recreational beach. Such seismic tests involved setting explosions to create small, artificial earth tremors. Soundwaves penetrate the earth and bounce back to a surface recording station at speeds that relate to the composition of the rock layers they strike below. Local residents were rumoured to be unimpressed with cracked windows from the shock waves, but the tests did reveal a gentle anticline underlying the metropolitan coastline. Beach Petroleum hurriedly set up a drill rig at Grange beach, 8 kilometres further north and bored down 600 metres, whereupon bedrock was struck and the rig just as quickly dismantled again.

Another hole had been spudded at Mount Salt, west of Mount Schank in the South East, but it too, was dry. A more promising prospect was Geltwood Beach, near Millicent in the same general vicinity. This bottom corner of the state was an area that had long fascinated Reg. It was here, at Alf Flat, east of Salt Creek, on the shores of the Coorong, that Australia's very first oil rig had been erected in search of the source of the elusive 'coastal bitumen' or coorongite that periodically washed up along the southern beaches.

In 1852, seven years before Colonel Edwin Drake first struck 'rock oil' at Titusville, Pennsylvania, USA, CW Stuart, leader of the 8th Victorian gold escort, had discovered a dark, rubbery, flammable substance lying about on the swampy ground at Alf Flat. When the area was taken up as a sheep run five years later, the shepherds cut strips of this material and burnt it in lamps. It was intriguing stuff. In some years it would accumulate at the edge of seasonal lagoons like a thick coat of paint; drying in large black sheets covering the mud or hanging in elastic, vile-looking strands from the reeds. Other years—typically the dry times—there would be none.

When the American oil rush began and Australia began importing kerosene as a source of illuminant, word soon spread to the Adelaide newspapers that a liquid fuel could be distilled from this local material, which soon became known as 'coorongite.' A ton of coorongite had yielded 120 gallons of crude oil, which had been further refined to produce 70 gallons of kerosene, 13 gallons of liquid paraffin and 7 gallons of black varnish. Many people concluded that the coorongite indicated a large source of petroleum at shallow depth. Others were certain that it was a vegetable product, not petroleum at all, and thereby fuelled the roots of a bitter debate that would last for decades.

In 1866, Australia's first oil exploration rig was erected at Alf Flat to plumb the source of this 'oil.' Drilling crawled along for two years until the well was finally declared dry at only six and a half metres. Various companies and individuals continued the search until the 1930s—all without success. By that time, scientists had confirmed that the coorongite was actually the dried residue of a freshwater algae, *Botryococcus braunii*, that blooms in the wet seasons and spreads over the surface of lagoons and ponds—an intriguing 'oil plant' comprising up to 30% hydrocarbons in the green phase of its growth cycle and up to 75% in its red phase.

The debate over the origins of coorongite had been confused by the fact that tar-like material cast up by the sea onto the beaches of southern Australian—the so-called 'coastal bitumen'—also collected on the shores of the Coorong. Pieces of this material had been known to reach the shore in a pliable, plastic state and also in great slabs weighing 50 kilograms or more. Similarities in appearance led to these two types of rubbery material—bitumen from the sea and coorongite from nearby freshwater lagoons—being grouped together.

Government geologist Arthur Wade in his five-month survey of, *The supposed oil-bearing areas of South Australia*, in 1915, tried to differentiate between the two types of material and at the same time dispel the myth that either would yield commercial oil deposits.

'I have found many fallacies with regard to petroleum to be rampant in South Australia, especially amongst the most enthusiastic,' he reported. 'The only regret which I feel in concluding my work is that I have been unable to treat the subject of this report with more optimism. The need of oil in the Empire is great, but not sufficiently great to justify financial waste in the search for it.'

He concluded that the South East represented the best—albeit 'faint'— possibility of harbouring real oil, on the basis of the coastal bitumen or 'asphaltum' coming out of the sea. He theorised that the petroleum material originated in ancient undersea sedimentary beds that had been disturbed in more recent times by the earth's faulting and rifting along the southern continental shelf.

'Most of the specimens of asphaltum found or examined by us have been sunbaked and waterworn, but very numerous cases have been brought before my notice in which the bituminous material has been washed ashore in a soft, viscous condition, deep brown in colour and retaining much volatile matter,' he wrote. 'The asphaltum has a wide distribution entirely confined to the coast and extending from the southern shores of Tasmania to the western limits of WA. The fact that it comes ashore at times in viscous condition suggests that it has not travelled any very great distance.'

Reg was intrigued with the whole coastal bitumen story. Since he had first studied the area as a government geologist in the Department of Mines he had heard stories of offerings of bitumen being flung up by the waves. He believed the locals when they said that the offerings increased after rough stormy seas that scoured the sea floor, or following a geological disturbance, like the earthquake that had hit the town of Beachport in 1948. He also believed, like Wade, that the source of the bitumen was nearby. Perhaps Beach could find oil at the beach after all.

At Geltwood Beach Reg collected his own samples of coastal bitumen. And so did his dad. Several hundred kilograms of the stuff. Local fishermen told Reg of having their anchor lines covered in oil. He compiled a record of all the reported sightings and arranged an early warning system to alert him to the first winter storm of the season, after which he would hasten to the appropriate beach and gather lumps of fresh bitumen sticky with oil. He had samples analysed by reputable petroleum laboratories, with favourable results. But still there were those who steadfastly dismissed the importance of the coastal bitumen and implied that Reg brought disrepute upon the whole industry by pursuing such a fanciful notion. When some technicians were forced to admit that the Geltwood samples were a genuine crude oil, they then dispensed doubt by pondering the likelihood of spillage from a passing oil tanker or some other inadvertent dumping of oil.

Reg was at a loss to explain such steadfast resistance. 'Frankly I am heartily sick of this buffoonery,' he confided in a letter to a colleague, 'the oils have been stranding on these coasts annually since white man first came to Australia—long before tankers were invented.'

'The crudes must be from a very extensive southern Australian oil province, immediately under and beyond the sea coast,' he wrote in November 1962 in a report to the Royal Society of South Australia—this, three years before Australia's huge Bass Strait oil fields were discovered.

Since there was such resistance to what seemed to him to be a siren call for an oil zone, Reg knew that a serious scientific investigation of the phenomenon would be up to him. This blinkered attitude was reminiscent of the scepticism he had first encountered when espousing the petroleum prospects of the Great Artesian Basin. But unlike in central Australia, here, in the southern waters, there was oil actually seeping out of the earth. What more could an oil man ask for? The only riddle remaining was to pin point exactly from where it was emanating and see if they could reach it with their drill.

Geosurveys geophysicists unspooled their seismic cables in the South East and identified a ridge of buried anticlines extending along the coast. Geltwood Beach, where the oil and bitumen came ashore in such bounty, was in the midst of all this geological faulting and folding. The Beach Petroleum directors agreed with Reg that this inaccessible piece of coast, named for the wreck of the steamer Geltwood that had foundered there in 1876, seemed to be a first-class drilling target. They hoped to have the site ready early in 1963.

For Reg, though, exploring from the shore was not enough. He wanted to look under the sea as well as along its edges. In typical Sprigg style he attacked the problem with gusto. He would build a boat he decided, a proper oceanographic research vessel from which he could investigate the sea floor in the same way that he probed the terrestrial surfaces—examining the contours, taking samples of the substrate, and recording the plants and animals that colonised each site. He knew, too, that it had once been a goal of Mawson's to explore Australia's oceans. In 1957 Mawson had led a deputation to Premier Playford with a request to form a South Australian Oceanographic Institute but Playford, like most politicians and geologists alike, was focused on oil and uranium at the time, and the idea floundered and eventually drowned.

Six years later, come Sprigg to the rescue.

He had first raised the idea with Griselda four years earlier in the wake of the excitement of spudding Santos's first oil well at Innamincka. His wife had been dubious at the thought of relinquishing the family nest egg to romantic dreams of the sea, but Reg was so darn convincing when he had an idea to explore. What's more he had his friend Dennis O'Malley urging him on. Dennis was Chief Engineer of the Department of Marine and Harbours. He had worked with Reg in the late 1940s when the state government had been investigating the possible construction of a deep water harbour at Robe in the South East and the two shared an interest in the structure of the South Australian coastline and sea floor investigations.

'You should see old Arno's boat,' Dennis had told Reg. 'She's a canoe-stern based on the MV *Atlantis*. You could build one the same. I'll throw in the mast and the keel. What do you think?'

Reg didn't need much persuasion, and now, four years and $160 000 later, she was almost finished.

IN SEARLES SHIPYARD at Birkenhead, Port Adelaide, Reg gazed at his pretty little wooden ketch in dry dock and began planning her adventures. He would bring the South Australian Oceanographic Institute to life and realise one of Mawson's goals. He also had his own dreams to fulfill. He hoped his new ship would finally solve the coastal bitumen mystery and above all, he hoped to lay to rest another controversy—findings from his own research, back in 1947 that had never been satisfactorily accepted by the local scientific community.

It was a saga reminiscent of his Ediacara fossils discovery that had happened just 12 months earlier, in March 1946. *They* had been largely scorned by the scientific establishment until amateur fossil hunters had harvested them by the truck-load and the experts rejoiced. In the same way, his discovery of giant canyons under the sea had been considered suspect when the Navy was unable to confirm his results a year later. With his own boat he hoped he could finally confirm his own findings and throw off the cloak of doubt once and for all.

It was a confounding internal examination for him, to reflect on that time, almost 20 years ago, when, within a year he had discovered the world's oldest animals and possibly its deepest ocean canyons. He had been celebrated for neither: Had in fact, been mocked and openly doubted. His natural inquisitiveness had motivated him then—as now; not the desire for celebrity. And yet he yearned for true acceptance. Some acknowledgement that his discoveries were profound and the product of thoughtful reasoning, not dumb luck. As a young man in his twenties the rejection of his work had been acute. Now, as a 43-year old, the wounds had healed, but the phantom pains still twinged occasionally, even after all that time.

IT WAS A foreboding day on the 26 April 1963, when MV *Saori*—her name an acronym of the South Australian Oceanographic Research Institute—rolled down the slips and into the water at 5.15 pm. There were comments among the assembled guests about the fortunes of a ship launched in such weather, as thunder, lightning and pouring rain drenched the boat from all sides. Reg and Griselda arrived late to the port after their car had stalled in a flooded roadside drain on the way to the ceremony, but they made it in time for Tom Playford's speech and for Griselda to dejectedly waste a whole bottle of perfectly good champagne on *Saori's* bow.

The winter months provided little opportunity to thoroughly test the new boat so it was the second weekend in August before she had her inaugural cruise. Naturally Reg selected an area south of Kangaroo Island over the edge of the continental shelf for her deep sea trials. This was the site of his River Murray submarine canyons and he was eager to prove their presence on the same voyage.

Saori had been carefully designed and fitted as a research vessel. In addition to all the usual equipment she had a shallow echo sounder and a deep, audio echo sounder ranging to 4000 metres—a rarity on non-naval

ROCK STAR

Geosurveys's oceanographic research vessel, 60-foot (18-metre), 85-ton MV *Saori*, launched 26 April 1963. The black and silver deep-sea diving chamber hangs from Saori's stern.

vessels at the time—a spacious map room and a glass viewing window in the bottom of the fish well. Sleeping berths and dining facilities could comfortably accommodate six on board.

Fifty kilometres south of Kangaroo Island they sighted Young Rocks—a series of granite knolls emerging 10 metres above the water. They were at about 90 metres deep and the granite rocks rose dramatically from the flat sandy bottom. Reg knew they were about halfway to the edge of the continental shelf where the sea floor would fall away sharply into the 5000-metre depths of the Jeffrey Deep. As soon as they crossed the shelf edge they recorded dense 'clouds' on the ecograph corresponding to enormous concentrations of fish and plankton that floated over the abyss at 350 metres during the day and rose to about 200 metres at night. Sharks appeared as blips near the sea floor on the shelf or at intermediate depth immediately beyond the shelf edge. Periodic successions of staccato signals were eventually recognised as whale sonar when a pod of about 30 whales (9- to 12-metres-long) homed on the vessel.

It was an oceanographer's paradise and Reg was enthralled. But for him, by far the most spectacular element of the cruise was when the echo sounder's audio signal changed from a sharp return echo bouncing back from the sea bottom to a whistling sound as the bottom fell away and the ship encountered the first of the undersea canyons—deep furrows in the steep slope that formed the edge of the continental shelf.

The ship took an east-west traverse following the edge of the shelf and recorded seven canyons plunging away to beyond the range of the echo sounder within a 60-kilometre-wide band. The heads of the canyons were at about the 75-fathom (137-metre) contour line, which explained how HMAS *Barcoo* had failed to record them in 1948, when it had made a traverse following about the 60 fathom-mark, five miles closer to shore than HMAS *Lachlan's* original voyage. *Barcoo's* failure to confirm *Lachlan's* results had discredited Reg's entire theory. Now, at last, his findings had been confirmed.

Some of the canyons were partly filled with debris and great landslide blocks, others were steep narrow gashes with a floor that could only be guessed at. Reg hypothesised that the apparent undersea slumping and 'landsliding' down the vertical canyon walls could be contributing to the escape of coastal bitumen if oil-bearing sands were severed in the process. It was an intriguing thought.

In fact the whole voyage was intriguing and it left Reg bursting for more. He wanted to get down there amongst it—actually feel the bottom with his fingers. Peering through the viewing window was simply not enough.

Reg's agile mind soon formulated the perfect solution. He would incorporate the new art of aqualung diving into Geosurveys's repertoire.

With the freedom granted by the innovative new self-contained underwater breathing apparatus (SCUBA) his geological staff could conduct a complete survey of the sea floor. *Saori*-assisted divers could take sediment samples and gravity measurements of the sea bottom, thereby fulfilling Geosurveys's contract to explore the Gulf St Vincent petroleum exploration lease area *and* enabling *Saori* to earn her keep. Reg's new passion for investigating the plants, animals and terrain of the marine world could be nourished at the same time. Almost nothing was known about life under the South Australian oceans. Sir Joseph Verco had dredged along the coast around the turn of the century, hauling up jumbled bags of material that had occupied taxonimists in the years since; but that was it. Nobody had observed or recorded life in situ.

For the remainder of 1963 the Simpson was forgotten—deserts replaced with oceans; red sand and spinifex exchanged for beach sand and seagrass.

Never one to tiptoe cautiously into a new venture, Reg proceeded at full pelt and proposed the idea of a diving chamber to complete the operation. A man secured inside an atmospheric chamber could descend to depths unattainable by a free diver. He would also avoid all the health risks and time-consuming decompression pauses associated with scuba diving.

It was another bold and daring Reg Sprigg plan. The modern art of diving with an aqualung was only 20 years old. No one had built a home-made diving chamber before. Taking gravity measurements on the sea floor was unheard of. And here was Reg proposing to combine all three. Yet no one at Geosurveys thought twice about it. This was the next job on the drawing board and they would follow their leader fearlessly forth.

'Reg said, "we're going to start this off-shore geo stuff and you're all going to be trained as divers",' recalls Clyde Thomson, who was an 18-year-old geological assistant with Geosurveys in his first job since leaving school. 'The general response was, "goodo",' he remembers. 'That was it. Reg; Darby—they just assumed we were doing it. And no one refused, because Reg used to gather around him people who were like-minded; who thought, okay, this'll be good fun.'

To learn how to scuba dive Reg made contact with Peter Warman and David Burchell. Peter was in charge of the South Australia Police Underwater Rescue Squad formed only six years earlier in 1957. Dave was a member of the fledgling Underwater Explorers' Club and the local agent for aqualung supplier Airdive Equipment. According to Dave, 'the number of aqualungs in South Australia at that time could have been counted on one foot.'

Dave was a determined, inspiring and fun-loving diver who often made feet the subject of his jokes. He had been just a lad when he met with an horrific accident riding the *Blue Lake Express* from Keith to Adelaide.

It was a hot summer day when Dave and his mate rode the train, sitting on the access platform between two carriages, their legs swinging over the side. When the train approached a wooden siding and the three railway gangers squatting in the shade of the stunted mallee scrub began waving in unison, the two boys waved back with enthusiasm. But the gangers had recognised the signs of tragedy racing toward them.

'Even as they sprang to their feet shouting unheard warnings, the sickening crack of the first boy's legs smashing into the siding steps, came to them. The boy's body—hand still raised as if in greeting—arced through the air and disappeared behind the gangers' shed. The second youth, knocked sideways by the impact of first his friend, and then the steps, involuntarily tightened his grip on the handrail of the carriage, and even the searing pain of being dragged the full length of the siding, over the rough weathered jarrah, did not loosen his grip.

'The first boy's body thumped into the loose cinders 20 feet away, its momentum carrying it on, ploughing and tearing until it ended up—like some sacrificial offering—at the feet of a woman returning from the store.'

The accident left Dave—the first boy—with only a stump of a right leg and 'one foot in the grave,' as he later named his memoir. The second boy survived with two broken legs and 42 large jarrah splinters in his backside.

After a stint in the Air Force and amateur success as a high platform diver, Dave had found his niche in the underwater world of aqualung diving. He organised dive schools at Port Noarlunga reef and various suburban swimming pools before constructing his own seven-metre-high, two-and-a-half-metre-diameter training tank, in which he trained divers from the police, the Harbours Board, Engineering and Water Supply, and other government departments.

'In its way the tank was an unparalleled success, and like *The Specialist* with his privies, when Don Mitchell and I got discouraged with life we would go out the back and just stare at it,' he writes.

Dave rated his Geosurveys association as one of his most interesting assignments.

'When Peter and I arranged a meeting to draw up a training programme for the Geos geologists, I was all set to go a bit easy on them. None of them were very fit and both Reg and Darby, were not exactly lads any more. But Peter wouldn't hear of it. "Be buggered," he said, "if they want to learn to dive, they'll get the full treatment, right from the start."'

Peter was a tough, efficient disciplinarian and Dave knew what the 'full treatment' entailed. He was already starting to feel sorry for his new pupils.

'Warman liked everything done on the double. During the training of the police candidates for his aqualung squad, the only time he would allow the poor characters to wear flippers was when they were running from one place to another.'

With a touch of compromise Dave and Peter drew up a comprehensive three-week programme covering both the practical and theoretical aspects of diving. The theory involved lectures on diving physiology, psychology and physics. The practicals ranged from snorkel training in a private pool; half-mile swims without gear in the open sea; mask clearing and free ascents in the tank; to the more serious techniques required for deeper dives and staging against decompression sickness, or the bends. They also threw in as many sudden and unheralded acts as they could imagine, in an effort to gauge each student's reaction to the unexpected.

Reg recorded one such surprise test during his first open sea trial. They were assembled at Port Noarlunga reef—Peter and Dave and a couple of other experienced divers, and the five Geosurveys trainees. Nine of the initial 'volunteers' had flunked the initial training so that only Reg and Darby;

Jonas Radus, Wilfred Stackler's son-in-law and Geos's most experienced gravimeter operator; and two others were left.

'Now don't get any fancy ideas,' Peter barked at them as they completed their final suiting up and checked their gear. 'The hardest part is still to come. You've only graduated from the bathtub. The reef will probably finish off the rest of you.'

The police sergeant harboured no sympathy for his charges. Overconfidence was a dangerous dive buddy and he wanted to ensure that his new recruits had a suitable respect for the underwater world they were about to enter.

One by one the divers scrambled off the reef and disappeared under the water. Then it was Reg's turn.

'Within seconds I knew this was my new world,' he remembers. 'Swimming down the seaward face of the block-strewn reef; probing in passing at the reef fishes; observing the beauty of the seaweeds and the colourful rock encrustations; all but blew my senses. I was transfixed by the beauty and deeply conscious of the coarse sand ripples, to the extent that I forgot what I was down about.'

The group descended to about 20 metres where Peter motioned for each trainee to pair with an instructor. Reg landed Peter. They swam on together, Peter at the rear.

'Next thing, to my unutterable horror, Peter ripped out my mouthpiece and shoved it out of reach into the back of my weight belt,' Reg records. 'Only one thing mattered—making it hell-bent for the surface. Halfway up I felt a tug on my legs and strong hands grabbed my feet, slowing me down as I held my breath. Peter was holding me back. I lashed out, sucked in a gut full of water and fought to free myself. Finally I broke lose—so I thought—but Peter had intentionally let go.'

Reg broke the surface coughing out blood and sea water. He struggled for breath with which to abuse his instructor, who was about to surface beside him, but Peter beat him to it.

'You brainless bastard, Sprigg!' Warman blasted. 'Call yourself a bloody scientist. Lucky your lungs are not protruding out of your useless trap at the speed you shot up.'

Suitably chastised, Reg was forced to listen in humiliated silence while Peter described his apparently recklessly speedy ascent to the other divers surfacing all around him. A hasty yet controlled rise to the surface, slowly exhaling air on the way to prevent over-inflation of the lungs, was the key to an emergency ascent in scuba diving. Reg had just failed miserably. He felt his enthusiasm for diving waning, but was instructed to descend before he'd even had a chance to compose himself. No sooner had they reached the bottom than Peter once again snatched his mouthpiece away.

'I started off slowly, remembering my last lesson all too vividly. Halfway up I was sure I wouldn't last the time distance. My lungs were craving for more air and I felt at the limit. I speeded up. But a touch on my leg brought me back. I made the supreme effort. In a few seconds Peter broke the surface alongside of me.'

'Good lad,' he commended the older novice diver in a fatherly way. 'I knew you could do it. Remember, as long as mates are around, it is better to drown…there is always a probability that you can be revived.'

It was an interesting philosophy that Reg hoped never to test.

Peter ruled with a mixture of threats and encouragement with plenty of scolding and bawlings-out thrown in. But Reg found it impossible not to like and admire him. After they had swum the kilometre back to shore Reg was surprised to hear him tell Griselda, 'Reg did a magnificent job. If he curbs his individualism he'll make the grade into deep diving.'

Peter took every opportunity to remind Reg that the confidence and geological single-mindedness that made him a success in the field could be a liability with 30 or 40 metres of water above his head. Darby and others who knew him well were likewise afraid that Reg would become too absorbed in his sampling and risk running out of air, or follow an interesting sea floor formation and lose his way back to the ship. Discipline and rigour could not be squandered in the deep sea environment.

WHILE SCUBA TRAINING was in progress, Darby also turned his attention to constructing the Geosurveys diving chamber. It had to be large enough to accommodate one, possibly two men and their equipment, and yet be small enough to be carried on board the 60-foot (18-metre), 85-ton MV *Saori*. While Darby pondered the physical and financial limitations at the Woodville workshop, Reg pondered the issue while staring out of a taxi window on the way from Sydney airport. In a metal dealer's yard he noted a number of large, rusting cylindrical boilers. He stopped and took the dimensions and later obtained the Boiler Inspector's clearance that such a vertical-standing, thick-walled boiler could be safely used at water pressures down to a maximum depth of 125 metres. It was ten times less than Reg was secretly aspiring to, but he agreed with Darby that a disused boiler would make an ideal diving chamber for *Saori*.

It was soon delivered to the Woodville depot and Darby began the modifications. He installed a round entry hatch at the top that could be opened and closed by two hand wheels, allowing it to be operated from both inside and out. In the half-inch-thick steel plate walls he cut three equally spaced two-inch-thick-glass port-hole windows at eye level so the chamber occupant could make visual observations of the sea bed and maintain contact with any accompanying scuba divers. He ran an iron ladder up the outside for the operator to climb in and out through the top, and attached releasable concrete weights to the bottom for negative buoyancy.

Inside there was a seat for the gravimeter operator, a chamber-to-ship telephone system, an emergency air supply and a carbon dioxide absorber in case of prolonged accidental submersion. It had a two-metre-diameter stabiliser at the base and was painted a distinctive black and silver checkerboard pattern with the Geosurveys logo on the side. When the three-metre-high, 4.5-ton chamber was complete, Darby made a waterproof housing to contain and protect the $20 000 gravity meter that would reside inside the chamber. He also designed and installed the triangular wooden structure that would harness the chamber over *Saori's* stern. The chamber could then be raised and lowered by two 4000-kilogram-breaking-strain nylon ropes, powered by compressed-air winches.

ROCK STAR

Darby was rightfully proud of the whole system and always took control of chamber operations when he was aboard ship as Dive Master. However, it was not a capsule for the nervous or claustrophobic. With internal dimensions of only one and a half metres high and one metre across, there was not enough room to stand up and the round walls pressed in on you with a stifling closeness. The thought of taking under-sea gravity

Geosurveys Operations Manager, Darby von Sanden, in full scuba gear, climbs into the diving chamber he designed and constructed.

184

measurements in this chamber was like contemplating being lowered to the sea floor in something the size of an old wooden wine barrel.

But the chamber's confines did not deter a man who had dived in the narrow blackness of flooded mine shafts, and as soon as Dave Burchell saw the chamber hanging from *Saori's* stern he was itching to try it out. His enthusiasm was fortunate because Geosurveys needed someone to test the emergency safety features of the chamber and not surprisingly no other volunteer had stepped forward for such a potentially risky job.

'I was busting for the chance to be first man down,' says Dave. 'As it turned out I needn't have worried. When starters were called I had no serious rivals for the honour. Peter wouldn't have a bar of it, and the others were really not experienced enough. The interesting part, as far as I was concerned, was not so much going down in the thing, but escaping from it when it was on the sea bed.'

Testing the dive chamber was slated as the last task in the official Geosurveys scuba-diving course. First the chamber would be checked for leaks, then Dave would perform an emergency escape. If all went well, the trainees would then each perform an escape in turn to earn their complete Geos scuba colours.

After the pressures of the diving course there was a festive air aboard ship on the day that *Saori* ploughed her way past the breakwaters at Outer Harbor. The entire Police Aqualung Squad was on board to view the spectacle, in addition to some diving mates of Dave's and some curious Geosurveys staffers. A crowd of heads leant over the side and others held off in the ship's rowboat as the chamber—with Dave inside in wetsuit, weight-belt and mask—disappeared under the surface. In a minute or two, the chamber settled on the bottom, 14 metres below.

When the cloud of sand and mud from the disturbance had cleared, Dave exchanged 'all okay' signals through the portholes with the scuba divers that had escorted him down and opened the two seacocks near the floor to begin flooding the chamber. Twin jets of water sprang through the valves with surprising force and in no time the floor was awash. Dave reached above his head to spin the control wheel of the hatch cover to the 'open' position and placed his hand on his scuba tank, reassuring himself that his air supply was readily accessible if he needed it in a hurry.

Except for the inrushing water sloshing over his feet, nothing happened for a few minutes, and Dave amused himself by writing rude messages on his diving slate to the divers outside his windows. As a diver, he found it an odd sensation, being underwater in a capsule—dry and breathing normal air; unencumbered by an air hose and scuba tank but imprisoned from the freedom of the water. As the water swirled up his thighs the pressure inside the chamber rose as the air inside was compressed by the water gushing in from the bottom. When the water reached his waist, Dave observed the curiosity of the depth gauge on his wrist registering a pressure reading of six metres. As the water crept up to his shoulders, water also began trickling down on him from above as it started leaking through the unlocked hatch cover. Not until the air pressure inside the chamber equalled the surrounding water pressure outside, would he be able to fully open the hatch and escape. It was nearly time.

'By now the water level inside the chamber was nearly up to my chin and, suppressing a mild wave of claustrophobia at the thought of the hatch being jammed, I took a couple of deep breaths and standing on the ledge that usually housed the gravity meter instruments, pushed against the hatch as hard as I could,' Dave recorded.

'The result was like standing under a waterfall. The heavy spring-loaded hatch, after a slow start, suddenly flew open and as the air in the chamber, compressed to about two and a half atmospheres, was released and took off for the surface, 14 metres of water pressure poured in.

'Waiting a second or two until the boil-over had subsided, I swam out the open hatch and free-ascended to the surface.'

He was grinning when his head broke the surface and there was applause from all the spectators.

'Magnificent Darby! It's terrific,' Dave pronounced.

He then took each of the Geosurveys divers down in turn and they practised flooding the chamber and using the emergency scuba tank to swim to the surface in case an escape had to be made in deeper water. It was like a passing-out parade as Reg and the four other new divers completed their final test and climbed elatedly back on board. *Saori* and her divers were finally ready for real marine exploration.

THE FIRST PROJECT to be tackled was a gravity survey of Gulf St Vincent—a shallow marine basin that separates Adelaide from the eastern shores of Yorke Peninsula. At its southern mouth the gulf is protected by Kangaroo Island, around which the water flows to the eastern Indian Ocean through the westerly channel of Investigator Strait or the rough, narrow, easterly channel of Backstairs Passage. Together, the Gulf and its approaches covered about 13 000 kilometres and Beach had petroleum exploration rights over the entire area and all of the peninsula. They had already conducted a gravity survey of southern Yorke Peninsula and recorded several gravity highs that could indicate oil field anticlines. Furthermore, traces of oil had been found in ancient marine sediments near Minlaton on the peninsula many years before and there was speculation that these sediments could extend and thicken beneath the gulf.

While Beach's first 'off-shore' drill rig was erected on a tidal sand bar on Troubridge Island five kilometres off the peninsula coast to investigate one of these gravity anomalies, Reg and his *Saori* team began crisscrossing the gulf with east-west traverses 10 to 15 kilometres apart, taking gravity readings on the sea bottom every couple of kilometres along the traverse line. They would survey the gulf in the same way they had measured the Simpson. At each measuring station the diving chamber would be lowered to the bottom with gravimeter operator Jonas Radus inside. Usually Reg or another scuba diver would accompany the chamber down and collect samples and make ecological observations for the 10 or 15 minutes it took Jonas to make his reading. Then chamber and diver would ascend and move to the next station.

The divers called it 'bounce' diving and they modelled their at-depth times on the only out-dated Navy dive tables they could find. They would start diving soon after daybreak in shallow 10-metre waters close to shore, make their way across the deepest section of the gulf down to 50 metres, and finish off with increasingly shallower dives. By always finishing with shallower dives, limiting each dive to less than 20 minutes, and ascending slowly they believed that long decompression times were not required, and they could manage 10 to 20 dives per day.

'With luck we could achieve a dive per hour for up to 15 hours a day,' Reg says.

Often the days didn't finish until well after nightfall.

It wasn't until later, after Reg experienced a mild case of decompression sickness or 'the bends,' when he went flying immediately after a dive, that he consulted with Hugh Le Messurier, a doctor of hyperbaric medicine at the Royal Adelaide Hospital. When Le Messurier reviewed Reg's diving records he was incredulous. Reg's daily accumulated times at depth meant that that he could have been hit by severe bends or even death on at least 13 occasions. There was a much better way to structure his dives, the doctor said.

Dr Le Messurier had visited Broome, at the top of Western Australia, where the Okinawan pearl divers had developed their own decompression times through trial and fatal error. Following deep dives, to 50 or 70 metres, they would start decompressions much deeper, pausing at 30 metres or so, then again at 10 metres, and five metres, and rising gradually to the surface.

Back in Adelaide Le Messurier and his PhD student Brian Hills tested the empirical decompression profiles of the Japanese divers using a hypobaric chamber at Adelaide University funded by the Department of Civil Aviation. They used goats as their subjects.

'Needless to say these animals often escaped and could usually be found devouring the best flower beds on campus,' Hills recalled.

'Unlike naval divers who are paid a fixed salary, pearl divers were paid according to the quantity of pearl shell which they harvested,' Hills explained. 'Thus they had a great incentive to minimize decompression time during which they were suspended in the ocean out of reach of the oyster beds...It was remarkable how they could decompress safely in half the time prescribed by the US Naval diving tables. The whole secret to their success lay in much deeper initial stops.'

'It wasn't the practice anywhere,' says Dr Scoresby Shepherd, a marine biologist who became *Saori's* primary diver, along with Reg. 'Le Messurier said it was a great idea—start decompressing much deeper than recommended in the tables and come up slowly. So Reg and I used to follow his advice and this may well have been the reason that we never had any problems.

'Those old dive tables were pretty primitive. They were developed on caissons in England in the 1920s. They weren't much practical use for us.

'But there were no rules anyway in those days,' he adds. 'I mean, we were doing ten or more dives a day. They were working to a pretty tight schedule. Today's dive computers would *never* let you do anything like that.'

In 1963, Scoresby was a lawyer with a passionate interest in marine biology. 'I had a consuming interest and a boring career,' he laughs, 'and you can see what took precedence ultimately,' he says from his desk at the South Australian Research and Development Institute's division of Aquatic Sciences. In 2006 he was made an Officer of the Order of Australia for his services to marine sciences. It was Reg Sprigg and *Saori* that gave him his real start in underwater research.

'I got to know Reg through the museum research group. I was an honorary at the museum and we were diving on weekends and collecting for the museum. I started off being interested in echinoderms—sea stars, brittle-stars and sea urchins—and then it expanded to everything.'

In December 1963, Reg invited Scoresby aboard *Saori* for a 10-day diving trip that Scoresby recorded in his dive log as the '*Saori* one expedition.' It was the first of many diving safaris that the two would make together.

'Reg was very enthusiastic, and he was great company,' says Scoresby. 'He would ring me up and say, "we're going out tomorrow, can you take a few days off?" And I would say…Well, I'd never say no. I'd take a few days off work—from the legal practice—and race down to meet them. Make up the time another time.

'Those were the very early days of diving so there was a lot of interest generally in what we might find. Most boats were there for professional purposes—line fishing. And there was nothing else at all. It was a complete blank. Then Reg came along.

'I know he had difficulty persuading his company to find the money for the boat. But he did. And I suppose he put in a lot of his own money, so the boat was built. He was keen to use it to foster marine research. He was always looking for something new.'

Besides—or perhaps because of—their shared infatuation with life on the sea bottom Reg and Scoresby had a similar approach to diving. They were both very independent divers who became completely engrossed in their collecting. In today's scuba world of dedicated dive buddies, where the welfare of your buddy is meant to be your primary concern, they would be considered poor partners.

'I never worried about anyone else,' admits Scoresby. 'And Reg was the same. So he was a good diver because we'd look after ourselves and then we'd look around and signal at the end and say, "I'm going up," and he'd probably come up too, or he'd say, "I'm staying here," and I'd go up.'

While some of the Geosurveys staff worried that such single-mindedness could lead Reg into trouble, in fact it was Scoresby that inadvertently caused the greatest alarm.

It was April 1964. *Saori* was making a gravity survey from Noarlunga, directly across to the heel of Yorke Peninsula, and back again. Reg had planned a mammoth operation—starting at sunrise and finishing at midnight—to complete the first Gulf St Vincent survey. Jonas was in the dive chamber, as usual, and Scoresby was one of a number of divers on the collecting team. At each gravity station Jonas was lowered into the water to take measurements and a diver accompanied him to record the ecology of the sea bottom and collect samples. By the time the boat reached the peninsula and was heading east for home, it was well into the afternoon. They had been at sea for more than eight hours and the job was only half done.

Sometimes—to spare the divers, who had each accumulated many hours underwater—Jonas would send a message on the chamber telephone saying that the marine environment appeared to be identical to what had been recorded at the previous station: don't bother sending a diver down. He would then complete the gravity reading, the chamber would be hauled up and they would move on to the next site.

Off Troubridge Island, approaching the centre of the gulf, Jonas phoned up to say, 'there's a magnificent reef here, lots of interesting things, send a diver down.'

Scoresby was already kitted up and eager—as ever—to see what was there. He threw over the shot line—a lump of lead tied to a rope that would settle on the bottom and lead the diver back to the boat—called out to the captain, Arno Mittler, 'I'm going down!' and jumped over the side.

'But Arno is a bit deaf you see, and he says he didn't hear me,' recalls Scoresby. 'Anyway, so I went down and made a collection, and then I was sitting on the bottom and I heard the boom, boom, boom as they were pulling the anchor up. I thought, that's strange, but I kept on working.

'Then I heard the motor start up and I thought, that's funny, so I came up to the surface. The sun was setting in the west and *Saori* was speeding off into the east.'

For anyone less at home in the sea, panic would have immediately descended.

Scoresby drew a tighter grip on his bag of specimens and sized up his situation. It was no use making a commotion, the ship was well out of range. With the sun now set he could see the lighthouse of Troubridge Island a few miles away and reckoned he could swim to shore. He dropped his weight belt and started swimming. He figured he'd make it by morning.

On board *Saori*, Darby began making preparations to anchor at the next gravity station a couple of kilometres away. It was then that he noticed the shot line over the side and began admonishing the divers for, 'Which bloody fool left that over!' Then, the obvious explanation hit him, 'Where's Scoresby?' he boomed.

Skipper Arno had the anchor up before it hit the bottom and was already steaming *Saori* back through her own wake. The sea was calm and clear but a strong tide was running south-west out of the gulf. Reg feared that Scoresby would be swept out into Investigator Strait.

As *Saori* surged forward at her top speed of 12 knots, all eyes peered into the watery gloom. Fifteen minutes had now passed since recognising the dilemma, and another fifteen prior to that since they had left the spot where Scoresby had gone over. As Dive Master, Darby felt the full weight of responsibility for Scoresby's abandonment. Somehow the instruction to limit diving to every alternate station had been misconstrued and the end result was a diver lost at sea in near total darkness. It was Darby's worst nightmare.

Then a call rang out: 'There he is!' And there was Scoresby, out to portside waving joyfully in the bright circle of *Saori's* mast light.

'I got a terrific bag of reef fauna,' he said, slinging an enormous bag of sponges over the side and climbing aboard.

Darby was speechless.

'Not to worry Darby,' Scoresby said. 'I figured that this tide would carry me out of Investigator Strait, the next one back, and so forth. It would only take two full exchanges of tide to bring around morning and I knew by then that you would be anchored here waiting for me. My main concern was to whether I could hang on to the bags of sponges!'

'It was just an oversight,' Scoresby says now, recalling this adventure more than 40 years later. 'But I can remember, absolutely clear to this day, the anxiety etched over all their faces…*peering,* side by side. Of course I yells out, "Where do you think you idiots have been!" ribbing them. The funny thing was, they never saw the joke. They were *deadly* serious, Reg; Darby. Darby was the most worried of the lot.'

While the Geosurveys seniors recomposed themselves and roundly chastised themselves for ever letting such a thing happen. Scoresby peeled off his wetsuit and serenely filled in his dive log:

'11 April 1964: Traverse from Troubridge Lighthouse due east. First dive two to three miles east of Troubridge Light. Here that Saori steamed off leaving me stranded, depth 25-30 feet. Bottom substratum travertine limestone, good growth of *Posidonia,* sponges (calcareous), numerous brittle-stars.'

The experience didn't deter him from further diving. His last dive that day was with Reg at 11 o'clock that night, 40 metres down, out in the centre of the gulf.

Parvancorina minchami

CHAPTER 9

SUCCESS IN THE 1960s

REG'S NEW MARINE toys consumed his energies for the summer of 1963-64 but he hadn't lost sight of the ultimate goal—the hunt for oil. In August, drilling had started at Geltwood Beach in the South East but by November the well had been plugged at 3750 metres and abandoned as a dry hole. Traces of gas were recorded but overall the £150 862 spent to drill the hole had yielded Beach nil. Reg was still convinced there was a source of oil in this Gambier/Otway Basin and continued his investigation of the elusive coastal bitumen. In August he had released a school of 50 drift bottles from *Saori* to determine where the tides would deposit them. Inside each bottle was a note requesting its return and promising a reward of 2 shillings 6 pence for those who did so. In January he released a second batch, in order to compare summer and winter water movements.

There was also quickening interest in the hydrocarbon potential of Australia's southern coastal waters a little further east, in the offshore Gippsland Basin. Australia's largest company, BHP, had just conducted an airborne magnetic survey over 29 000 kilometres of ocean to form a basic outline of the sedimentary basins in Bass Strait, and was following up with a seismic survey over their permit areas in the Gippsland and neighbouring Otway Basins. The traditional iron and steel company knew very little about petroleum exploration, but was studiously following the advice of American oil consultant Lewis Weeks. Weeks had visited Australia at BHP's request in 1960 and had pointed to the waters of Bass Strait as being the most prospective, discounting the Sydney Basin in which BHP had been most interested at the time.

Weeks was already an acquaintance of Reg's so during one of his trips to Australia, Reg invited him to South Australia to view the progress of his own company, Beach Petroleum, in the Otway Basin. They toured the South East together and Reg told Weeks the long story of the coastal bitumen controversy. As they stood on the beach at Cape Northumberland, staring out to sea, Weeks agreed he found the established resistance to the idea of a local origin for the oil seep, hard to fathom. Surely the bitumous deposits that Reg described were exactly the signs that oilmen sought in their search for oil.

As the two geologists walked the sand, discussing the tantalising idea of an oil province off the South East coast, they noticed small blobs of oil—about the size of a 5 cent piece—arriving on the beach. Weeks immediately took his hand-lens from his pocket and squatted down to inspect them.

'They're still gassing,' he announced. 'This is of very local origin.'

Reg was impressed. He had collected oily samples like this many times but had never thought to inspect them under a magnifying glass. Reg believed him when Weeks pronounced, 'One day Bass Strait will prove to be a major oil province.'

Although Reg was lured by the thought of oil under the sea he had never lost faith in the hydrocarbon potential of the vast Great Artesian Basin. And

SUCCESS IN THE 1960s

then, on New Year's Eve 1963, it had finally happened. Santos struck gas in the Gidgealpa-2 well…

Austrian geologist Heli Wopfner had by then fulfilled his four-year contract with Geosurveys and was working for the Department of Mines. He hadn't intended leaving Geos but when Reg asked him in mid-1960 to go up to the nickel venture at Mount Davies, 'just for a couple of weeks,' an alarm bell went off in Heli's head. He knew what Reg's 'just for a couple of weeks,' meant and he knew that he would likely be stuck up in the northwest corner of the state for at least a couple of months. For the sake of his family he couldn't accept this new assignment. The year before—1959—when Heli had been the Geosurveys/Santos on-site representative at the Innamincka well he had spent 10 months in the field.

'It got to the point where my kids wouldn't hardly know me when I got back home,' Heli says 'Something had to be done. Either my family would break up or I would break up. I had to have a change in environment.'

It was serendipitous that an opening at the department had come at the same time. Now, after three years, he was senior geologist of the newly created Petroleum Geology section.

After the frenetic excitement that had followed Australia's first oil discovery at Rough Range in late 1953, the industry had largely slid into the doldrums. Twenty-seven wells had been drilled in the Rough Range area over the following decade and all had proved disappointing. Even the original discovery pool—estimated at 70 000 barrels—was declared too small to be commercial and the famous well was eventually abandoned.

The naysayers, who had retreated one step after the Christmas 1953 discovery, had begun rehearsing their 'I told you so,' speeches when in late 1961 another Australian/American partnership struck oil at Moonie, in south-eastern Queensland's Surat Basin, not far from the very first gas finds at Roma. The 1765-barrel per day flow rate led to Australia's first commercial oil field and injected much-needed enthusiasm into the industry.

For Santos, after the fruitless drilling of the Betoota-1 well, which had followed the inaugural Innamincka well in 1959, there had been a lull in the Great Artesian Basin exploration programme. The Frome Broken Hill

Group (Frobilco) had once again lost faith in the region and pulled out of the partnership. Santos and Delhi continued. Five more wells were drilled but funds were drying up.

In 1963 the French Petroleum Company (Total Oil) joined Santos and Delhi in a farm-in agreement and Bonython and the Delhi directors selected the Gidgealpa anticline 70 kilometres south-west of Innamincka as their next drilling target. The companies' prior experience in the area had by now shown that some of the spectacularly large anticlines that Reg had discovered had the tendency to be 'bald-headed.' The layers of sediments thinned at the crests of the mounds to a point where they were unlikely to harbour hydrocarbons. This time Santos and Delhi decided to deliberately drill into the eastern flank of the Gidgealpa structure.

The well was spudded in August and had reached a total depth of 4000 metres by the end of November. With no oil shows, the company directors became dispirited and called for a six-month moratorium on further drilling. But there were others who felt that the Gidgealpa well contained revealing clues that needed to be pursued. The 65- to 250-million year old Mesozoic sediment layer that the drill had intersected was the thickest yet encountered in this part of the Great Artesian Basin, and directly underneath the Mesozoic sequence was an even older layer of Permian sands that had multiple gas shows. Hydrocarbons in 250–300 million year old Permian sediments was completely unexpected and the joint venture geologists largely ignored these indications as they drilled down deeper to where they believed any oil or gas would be found.

When Heli Wopfner, in his new position at the Department of Mines, inspected the logs from the well, he suspected that the Permian sands would, in fact, be prospective and he also suspected that they would stretch over the entire anticline in a thin blanket. Using the powers vested in him under the State Government Petroleum Act he gave the operators a choice of two options: Either adequately test the Permian sands in the Gidgealpa well—which he knew to be nigh impossible due to technical problems in the well—or drill a second hole at the top of the anticline.

SUCCESS IN THE 1960s

'Of course this started a really extensive controversy between geophysicists, between geologists, and between the various companies,' says Heli. 'Santos and Total didn't want to have a bar of a second well. They didn't want to test either, because they knew they couldn't. Delhi was divided. The chief geophysicist of Delhi resigned from Delhi because he was of the opinion that the structure would be bald-headed.

'It was great fun. The department made it quite clear we would not budge from our position—either one of the works would have to be carried out. It went to Tom Playford and it took Playford's well-known gentle persuasion and arm twisting with some of the directors of Santos, before Santos would finally agree that the second well would be drilled.'

The Gidgealpa-2 well was spudded on the 12 December and was monitored by the department every step of the way.

'Three senior geologists were sitting at the well when the critical section was to be drilled,' says Heli. 'On the 31 December 1963 at six o'clock in the morning we opened the test tools for the first drill stem test and obtained the results of 2.8 million cubic feet of gas. And that was the discovery!

'We immediately called to Delhi. They recalled the plane, which had already been on the way up, and Delhi loaded—I don't know how many dozen—champagne and white wine and so on, onto the aircraft so we could—in addition to the normal New Year's Eve celebration—also celebrate the discovery. And believe me…it was celebrated.'

It was South Australia's first petroleum discovery.

Although he was no longer a direct part of it, Reg Sprigg's faith in the Great Artesian Basin had been vindicated.

LIKE OILMEN AROUND the country Reg felt invigorated by the news of Moonie and Gidgealpa and the tantalising prospect of more petroleum to be found. His thoughts turned again to the interior of the continent and he took his family into the desert once more, crossing the Simpson this time from south to north—Marree to Alice Springs—scouting an alternative access route in the event of future oil discovery in the desert. Compagnie Generale de Geophysique were at that very moment completing their seismic track—soon to be nicknamed The French Line—west to east across the Simpson from Dalhousie to Birdsville, but Reg professed to be interested in finding a route between the dunes that could serve as a north-south corridor.

The four Spriggs were packed across the front seat of the Toyota, as usual; Margaret now 12 and Douglas about to turn 10. For the children, these outback adventures were precious. It was the only time they had their father all to themselves. They basked in his proximity and eagerly soaked up his numerous spontaneous lectures on geological features they encountered along the way, explorers who had traversed the area before them, and the uniqueness of the plants and animals that made the desert their home. Margaret delighted in having her father catch lizards with her while young Douglas probed the theories of aviation, astronomy and mechanics while he and his father changed tyres and mended punctures together.

'This is one of the finest things about a desert,' Griselda recorded in her diary. 'Father and son become very close…Reg and Douglas spent some time mending not one, but four punctures, and all the while Douglas discussed astronomy with his Dad. At home on many occasions Douglas badly wants to discuss things with his Dad, but I never yet remember one occasion when the telephone has not aborted their talks.'

In contrast to their last Simpson trip, the Spriggs would be a lone car this time—no backup except for a single fuel drop near Geosurveys Hill, an isolated outcrop in the middle of the desert that only just peaked above the level of the surrounding sand dunes.

It was another successful crossing. Along the way they crossed the path of the Caterpillar D9 blade that was scraping out the French Line; survived a 110 kilometres per hour sandstorm; and Reg named a distinctive H-shaped salt lake, Lake Griselda.

'The lake was finally measured at 31.3 miles long and about three quarters of a mile wide—a long skinny lake with a thin hard crust which broke down on pressure,' Griselda recorded. 'Those who know me personally will see the resemblance.'

She was also impressed with what she saw of the new French Line.

SUCCESS IN THE 1960s

'The French tracks, for which they had apologised, were as good as any graded tracks I have seen and were a tremendous help. They were put in possibly three to four months ago by their seismic teams. One reads often of the fine work done by the early pioneers opening up the country, but surely the geologists and geophysicists and their many assistants must be the modern pioneers: opening up hundreds of thousands of square miles of previously unknown territory. The heavy equipment required on site for a drill to operate makes it necessary for these roads to precede all operation on the drilling side. I only wish just a small percentage of the annoying people who so frequently ring up at home and ask "when is your husband going to find oil?" could just see the massive preparations which have to be made before any thought of drilling can be made.'

For Griselda, too, these family jaunts into the desert—despite the strain of packing and organising, which largely fell to her—were cherished escapes from the myriad daily demands on her husband's time.

A year later, in the September school holidays of 1965, the family crossed the continent from Eyre on the Great Australian Bight to Wyndham in the eastern Kimberleys, traversing the Great Victoria, Gibson and Great Sandy Deserts on the way. Foraging in the desert was interspersed with forays to the bottom of the sea, and the months hurtled by at the manic pace that Reg revelled in.

When he was home, working from Geosurveys's city office in Adelaide's Grenfell Street, there were always myriad reports and papers to prepare. In the mid-1960s the job of deciphering Reg's hurried scrawl and typing them all up fell to Dorothy Francis.

'I was a high speed shorthand typist. But Mr Sprigg didn't like dictation. He liked to scribble all his stuff out,' she remembers. 'It was all highly technical and written in a hurry. His writing was so appalling at times.

'One time, I'd finished a report all except one sentence. I pored over it and tried to work it out. I knew it was important. I didn't like to disturb him but eventually I went in and said, "I really can't read this sentence." He looked at it for a while and then he said, "Well, I don't know. I can't read it myself."'

'Working for Geosurveys was not the usual office job. It was a joy to work there. And he was such a nice person to work for. He'd never come out and just plonk stuff on your desk. He took the trouble to come and explain things to you…the difference between the different strata; the Cretaceous Period and so on. He'd give you books to read…"Well, read up on it," he'd say.

'He lent me lots of books to read. Not only was the work intelligent and different he *made* it interesting. He had the knack of inspiring you and arousing your curiosity to know more.

'When he was going away, he'd get everything ready in the Board Room. He'd give you his wallet—full of money—and a great list, and organise a taxi for you. And I went round to all these places and collected all this stuff. I'd pile it up in the Board room, all on the table and on the floor. He was extremely trusting. Fancy giving you his wallet full of money! The last thing you'd want to do was betray that trust.

'That's when he lost his favourite sun hat—the one he always liked to take into the field. And we got down to serious hunting and found it at the bottom of his wastepaper basket! He was a bit absent-minded at times. Every now and then there'd be a panic on and we'd all go flying around looking for stuff. He had so many big things on his mind that the small things were the ones that got forgotten. Like when Mrs Sprigg rang up and said that he'd forgotten to wear his socks. She said, take the money out of petty cash and go round to David Jones. She told me the size. When I got back I went into his office.

'"Excuse me Mr Sprigg," I said, "I think you'll need these."

'"Good Lord!" he said, when he looked down at his feet,' Dorothy laughs. 'His mind was full of much more important things.

'He used to work at a high speed. He was hardworking and expected you to keep up—there was no slacking off. But he was reasonable, too. He was always very nice to students et cetera who came for help or work experience or something. He never brushed them off in an unkind way. He was a kindly person. And always wanted the best of everything. He wanted the best presentation for sending all his work overseas—lovely vellum paper, leather bindings, everything the best you could possibly do. It was a joy to work there and use all the good things. Other bosses would say, no you can't have a piece of fresh carbon paper… He was a very nice person to work for.

'Then he'd go off again. The office would close down when he was away and I'd go home, or the agency I worked for would send me somewhere else until he came back. I always looked forward to when they'd call and say Mr Sprigg's back.'

IN THE SECOND half of the 1960s, Reg had multiple projects on the go, as usual. In addition to *Saori's* marine investigations and Beach's expanding operations, the nickel venture in South Australia's north-west—for which he was a company director and Geosurveys was technical consultant—continued. The original exploration area was now focused on a 130-square-kilometre nickeliferous ochre deposit at Wingellina, just over the border in Western Australia. The option of an open-cut mine to harvest the deposit, which had an average nickel content of 1% to 1.4%, was being actively investigated but the expense, combined with the huge amounts of water and energy required to treat the nickel kept the enthusiasm of their major shareholder—The International Nickel Company—restrained.

Reg's interests in the uranium deposits near Darwin, however, had disintegrated. In 1957 the Adelaide River mine, operated by Uranium Development and Prospecting—for whom he consulted—and its associated company Australian Uranium Corporation, had finally fallen victim to poor management, high overheads at the mine site, squabbling with government representatives, and a world-wide over-supply of uranium. It marked the end of Reg's final official association with the charismatic mineral that had launched his geological career.

In fact, uranium was losing its allure for politicians and investors alike as the Commonwealth government sales contracts with the UK/US Combined Development Agency (CDA) expired, and the British began sourcing their supply from Canada and the Congo instead. Although yellowcake was still being stockpiled at the Rum Jungle mine south of Darwin, the CDA contract had finished in 1962 and no new buyer had been found. The Radium Hill mine in the north-east of South Australia had also closed, in December 1961, only seven years after its official opening.

Despite its short lifetime the Radium Hill project had been a financial success for the state government. Within the first 18 months all operating costs had been recovered, full interest charges met and 16% of the $7 million capital investment in Radium Hill and the Port Pirie processing plant had been recovered. In its seven years of full operation the mine had provided more than $26 million net in overseas exchange and employment for up to 700 people.

The decision to close the mine had come as a shock to the inhabitants, however, when the news was announced only two months before the closure.

'Who killed Radium Hill?' enquired a message painted on the side of a black coffin being pulled along in the town's final annual Labour Day sports parade. 'I, said Tom, with my little bomb.'

In the decade of the town's existence, homes and lives had been established in the remote outback community. One hundred and seventeen babies had been born at the Australian Inland Mission Hospital and 12 young graves occupied the little cemetery. Facilities such as the concrete swimming pool, complete with kiosk and amenities, which was installed by the mines department at a cost of approximately £10 000 in the summer of 1956/57, had had only five years in which to be enjoyed. Likewise the Catholic Church, completed in 1956—the sole brick building in town. The resident parish priest, Father Vincent Shiel, a qualified architect and skilled tradesman, drew up plans for a building with 9-inch cavity walls and 10-foot ceilings, concrete floors and an iron roof, to house a church complete with two confessionals and a presbytery. He then made 42 000 cement bricks with which to construct it. A true labour of devotion.

In 1962 when the town and mine site were dismantled, buildings and equipment were redistributed throughout the country. The aim was to return the area to as near to its natural state as possible. What could not be salvaged would be demolished. The Electricity Trust of South Australia

SUCCESS IN THE 1960s

subsidised the removal of the wooden inter-denominational Protestant Church to the coal mine township of Leigh Creek. Hospital equipment was sent to the Australian Inland Mission hospital at Kununurra in north-eastern Western Australia. The two army Nissen huts that comprised the wet canteen became the carpenters' shop at Adelaide's Yatala prison. Even the land itself, was eventually returned to the lease-holders of Oulnina station upon payment of the original purchase sum of £750 in addition to £1250 for the improved water supply. Structures that couldn't be salvaged were smashed and shovelled into the ground. Mine trucks and miles of underground railway and pipelines were simply left where they were, entombed forever inside the sealed mine.

When it came time to demolish Father Shiel's brick church the wide concrete lintel above the entrance crashed down on to the bulldozer upon the first touch, smashing the machine's radiator and engine. The operator interpreted this as divine intervention and refused to push over a single brick. In 2007 the time-worn ruin still stands. The only remaining building in a township that once boasted 1000 souls.

REG'S DEPARTURE FROM the field of uranium mining did not leave him with spare time on his hands. With the viscous attraction of mercury, other projects flowed easily into the void.

Beach's gravity survey of the Adelaide coastal waters had found no obvious drilling targets so Geosurveys entered the new field of marine seismic, in typical poor-man fashion. Seismic data would give a more detailed picture of the sub-surface geology than gravity readings could provide. Unlike the two specialised Italian vessels that had conducted the seismic survey of Bass Strait for BHP, *Saori* kitted up with a length of land-based heavy steel seismic cable converted to a kilometre-long floating cable and geophone array manufactured by Darby in the Woodville workshop. The cable was wound onto an enormous drum at the back of the boat and radar was fitted to the ship for more accurate positioning. The Geosurveys geophysicists transferred their traditional recording gear on board and, so equipped, *Saori* set off into the waters of St Vincent Gulf and eventually into Cook Strait, New Zealand. Accompanied by a colourful assemblage of semaphore signalling flags aboard ship to warn other vessels of the danger, the seismic cable would be spooled out behind and massive explosions detonated to create small, artificial earth tremors 3 kilometres below the sea surface. The resultant 100-metre-high water fountains made *Saori* appear a mere cork bobbing in the water.

By the end of 1967 marine seismic surveys had proliferated from a rare oddity to such common practice that the department of shipping and transport was releasing warnings to mariners and fishermen about the danger of unexploded charges dropped during seismic surveys. Five tons of unexploded ammonium nitrate explosives had been found on the sea floor about 3 kilometres off the Victorian coast, consisting of one group of 60 to 70 charges—each weighing 45 kilograms—uniformly distributed over a distance of 130 kilometres between Golden Beach and Lakes Entrance, and another group of 1400 kilograms of explosives in the same vicinity.

By then, *Saori's* marine survey days were almost over. The seismic data off the coast of Adelaide had shown that the younger sediments in the St Vincent Basin were too shallow to harbour oil and the older Cambrian sediments too faulted and folded to be of serious petroleum interest. Beach, by then, was half owned by two Melbourne-based companies, North Broken Hill and South Broken Hill—the Collins House group, which had taken a 50% share placement in 1964. The group began steering Reg's company into more conservative directions and nixed some of his more outlandish plans, like the construction of an underwater dwelling for extensive submarine research of…say…the artificial reef he had built in 20 metres of water off the Yorke Peninsula.

'They'd loaded up hundreds and hundreds of bessa blocks,' remembers Scoresby Shepherd, 'and about 10 kilometres off Stansbury we built an artificial reef. This was typically Reg. He'd have an idea—"Let's build an artificial reef! And we'll make several configurations, just for fun, and one day we'll come back and see what it looks like." So we dumped all this stuff on the bottom from *Saori* and then went down and built all these walls and structures. We worked for hours and hours. And that was that. I've never gone back to see.'

On weekends Scoresby also helped Reg and Darby build the underwater living chamber—fashioned from another disused boiler 6 to 8 metres long with a diameter of 2 to 3 metres. It was never completed.

Other projects, like salt harvesting and experimental oyster farming at Coobowie on the Yorke Peninsula, were more private endeavours that largely fell outside of Beach's jurisdiction, yet still fed Reg's voracious appetite for activity and endeavour. He certainly had not lost his fascination with scuba diving. Since first seeing the smooth slope of the continental shelf fall away on *Saori's* echo sounder during her first deep sea trial in 1963, Reg had wanted to see the deep continental shelf for himself. He knew that it was impossible to plumb the depths he really wanted to explore but he was determined to reach the lower limits possible for a scuba diver breathing standard compressed air.

In the autumn of 1966 he had finally had his chance. He, Darby and David Burchell—also eager for the opportunity to make a deep dive in South Australia's relatively shallow coastal waters—boarded *Saori* in Robe and headed 20 miles out to sea south-west of Beachport. They were headed for a depth of about 80 metres at a spot Reg called 40-fathom beach. It was the site of what he believed to be an old Pleistocene, low sea level, erosional beach that he was most anxious to examine. By the time captain Arno Mittler dropped anchor, Reg had checked his gear and was already shrugging on his scuba tank.

Despite their excitement, each diver knew that this was a dive not to be taken lightly. Nitrogen narcosis or 'rapture of the deep'—a kind of drunken disembodiment where one loses one's senses and one's grip on reality—was almost a certainty at these depths. Running out of air was another possibility and suffering 'the bends' or decompression sickness was a risk on ascent. To avert all these potential problems the trio had taken careful precautions. Three spare scuba tanks had been secured to the shot line 45 metres down for anyone—most likely Reg—who ran out of air, and two additional breathing hoses tethered to the ship had been left suspended at five metres down in case emergency decompression treatment was necessary.

While they sat side-by-side at the gap in the deck rail with their legs hanging over the side, Dave also administered what he called 'a pep talk' on Reg's responsibility to himself and to the others.

'I don't think he appreciated it much,' Dave recalled, 'as he told me quite briefly and concisely when I'd finished to "go to hell"'.

'My job was to "ride shotgun" on the other two,' says Dave, 'and whilst I was not worried about Darby—for he keeps his mind on the job—I was a bit concerned about Reg. The old Sprigg, like most truly dedicated professional types, gets so completely carried away and lost in his geological fossicking that I'm sure he forgets where he is. It makes no difference if he is underwater or in the middle of the Simpson Desert, his environment means nothing to him.

'To add to my problems, Reg was an air sucker, and without any trouble at all when we dived to 63 metres [the day before], he all but flattened his cylinder, whilst Darby and I came up with ours still two-thirds full.'

Dave reiterated that the plan for this dive was for Reg to go first, then Darby, and then himself. They were to keep visual contact with each other and physical contact with the shot line at all times. As they were not wearing weight belts—since their wetsuits would lose buoyancy anyway due to the extreme pressure—they were to descend quickly using the shot line to pull themselves clear of the plunging ship. They were to stay two minutes on the bottom and come straight up.

Placing his hands over his face to protect his mask, Reg slipped feet first over the side. Darby followed, and then Dave. Almost at once the plan went awry.

Like the other two, as soon as Dave hit the water he spun around and started pulling hand-over-hand on the white nylon rope of the shot line as hard as he could. But with Reg and Darby hauling themselves down below him, combined with the depth and a relatively fast-running tide, Dave was receiving only handfuls of slack line. He was still bobbing helplessly at the surface when *Saori's* hull crashed down on the back of his neck, almost

SUCCESS IN THE 1960s

knocking him out. In desperation he heaved on the line and managed to move down about 3 metres as the ship was poised to deliver her next crushing blow. While his head cleared Dave felt around behind him, as best he could, in an effort to check that his regulator had not been dislodged from the cylinder valve. He knew he should really ascend and check his gear thoroughly but it would mean missing the dive and leaving Reg and Darby—oblivious to his troubles—on their own.

'Not bloody likely,' he thought, as he flipped upside down and recommenced pulling himself down the line for all he was worth.

Although late in the day, the water was a beautiful royal blue and the white shot line seemed to stretch on into infinity. It wasn't until he was 60 metres below the surface that he began to make out the shadows of the sea floor, and a further 6 metres or so before he could decipher the tiny figures of Reg and Darby on the bottom.

Reg glanced upwards to see Dave and the miniature *Saori* far above silhouetted against the setting sun, but wasted no time in his exploration of the sea bottom. It was typical Mount Gambier Tertiary limestone with serpentine solution channels a half a metre wide and 20 to 30 centimetres deep. He concentrated hard to absorb everything he was seeing and resist the creeping affects of narcosis.

'In a growing trance I told myself what I was seeing—talking to myself as though half of me was out of me, so to speak…note the structure…see the solitary corals…there is coarse shell sand, some cockles and plenty of basket polyzoa and little else.

'I then lay in a channel. It just accommodated my body. My mind was groggy now and I would have liked to remain, but Darby arrived at my side, pointed to his watch and then up.'

Their two minutes on the bottom was over.

As Dave floated down to join them, the three exchanged 'I'm okay' signs and Darby and Reg began their ascent.

It was Dave's turn to take in the moment. He held onto the shot line with one hand and gazed around at the unusual scene. It was a marine meadow with soft corals, sponges, sea squirts and lamp shells carpeting the sandy white bottom. He too, made a concentrated effort to memorise pertinent details, such as the amount of rock and marine growth per square yard, types of coral and fish life.

Back on top five minutes later, they compared notes and impressions with boyish enthusiasm. Reg was envious of the cabbage-sized lace coral Dave had souvenired but was equally as delighted with the rocks that he and Darby had sampled. Dave found them unattractive and completely meaningless but they had Reg really jumping. The geological structure of the sea bottom—sand-filled gutters and lanes in the limestone—told him that the surface at this spot had once been dry. He had lain on a sandy bed more than 80 metres under the sea, that had once been a dusty coastal plain during the peak of the last Ice Age 18 000 years ago. It had been an exciting dive indeed.

Dave too, was pleased with the day's result. Bounce diving to 82 metres without decompression was quite an achievement.

When he arrived back home, his wife, Ona, looked up from peeling potatoes in the kitchen to enquire after his dive.

'What was it like?' she asked, drying her hands.

'Just like a garden,' Dave sighed, the memory of it still fresh in his mind, 'just like a beautiful garden.'

Ona, acting like any wife whose husband completely lacks horticultural interests, couldn't let the opportunity go by, Dave recalled.

'Is that so,' she said, as dry as you like, 'just like a garden? I wouldn't have thought you could recognize one.'

By the late 1960s Reg's fortunes were blossoming. Although neither Geosurveys nor Beach had found oil of their own, both of his companies were prospering and the industry itself was buoyant.

Santos had followed up its Gidgealpa success with the discovery of an even bigger gas field at Moomba and the company had signed a contract to pipe gas from this major hydrocarbon province—now named the Cooper Basin—to the city of Adelaide. WAPET, who had started the oil boom back in 1953, had finally discovered a commercial oil field on remote

SUCCESS IN THE 1960s

Barrow Island, off the north-west coast of Western Australia; and BHP in partnership with Esso had discovered the mammoth Bass Strait oil and gas fields. By the end of the decade Australia would be 70% self-sufficient in oil, and natural gas would be flowing to Adelaide and Brisbane, with Perth, Melbourne and Sydney soon thereafter. The decade of the Pill, the Beatles, and the Vietnam War had borne witness to the country's most successful petroleum exploration period.

In late 1969 the country was also in the grip of an hysterical nickel boom. In the previous three or four years, while nickel-consuming stainless steel production had expanded rapidly around the world, labour disputes between the giant International Nickel Company (Inco) and its Canadian miners had seen supplies become increasingly scarce. It was perfect timing for Western Mining Corporation to bring Australia's first nickel mine into production at Kambalda, 50 kilometres south of Kalgoorlie in Western Australia. Its first consignment of concentrate was shipped overseas in August 1967 and as the company's output increased, so did the value of its shares. A second commercial nickel discovery was made by another gold-mining company at Scotia, 60 kilometres north of Kalgoorlie, at about the same time, and the boom was born.

It was two years later, in September 1969 when the small Adelaide-based exploration company that would eponymously give its name to the boom, entered the fray. Poseidon owned a nickel lease at Mount Windarra, 40 kilometres north of Kalgoorlie. Two days after the company started its first drill hole the value of its shares jumped suddenly from 33 cents to $1.48 in a day of brisk trading. When the company announced fantastic nickel grades of 3.5%, its stock soared to unimaginable heights in a four-month trading frenzy that took all the other oil and mining stocks with it.

One month into the Poseidon boom, the Sydney Stock Exchange was handling three times the daily volume of trades of the New York Exchange. The combined value of mining and oil shares traded on the Sydney Stock Exchange in August 1969 was $28 million. By December, a mere three months later, it was just below $200 million. This spectacular rise contributed to the value of mining and oil stocks traded on the Sydney

Stock Exchange exceeding $1 billion for the first year ever. During morning trading on Christmas Eve, more than 15.5 million shares changed hands in Sydney, compared to only 2 million on the same day the year before.

Poseidon stock had risen from 33 cents in late September, to $48 on the first trading day in December. And it kept on going. On New Year's Eve it rose another $40 to smash through the $200 mark and close the year at $210. Shareholders were jubilant. Traders were awestruck. The Poseidon boom was the wildest phenomenon ever recorded on Australian stock exchanges.

Rumours and front bar gossip from the Western Australia nickel fields were dispensed with lightning speed to stockbrokers all across the country. Telegraph operators were suspected of delaying telegrams while they updated their share portfolio based on confidential information relayed in transit. More than 250 claims were lodged in one week with the Mining Registrar of Leonora, the nearest town to Poseidon's lease area. Stockbrokers were packed three or four deep on the floor of the Sydney exchange in a crush that financial journalist Trevor Sykes described as akin to 'being in a rugby scrum but not as comfortable.' When one broker told his client on the phone 'we're all in turmoil here,' the client responded: 'Buy me some too,' not wishing to miss out on a new oil company share float. When a newspaper ran a story on mining claims and used a fictitious company, 'Fickle Nickel No Liability' to illustrate a point, a reader rang to ask where he could buy shares.

It was one crazy ride.

Although Reg's company, Nickel Mines of Australia Ltd, was not listed on the exchange and the joint venture with Inco had no startling news to announce, he was swept up in the fever nevertheless. A photo of Reg appeared in Melbourne's *The Age* newspaper on the 16 January 1970, under the headline: 'Another five men make their "paper" millions (and in just a few minutes).'

'Australia's caviar and champagne class, its millionaire population, must be growing at an enormous pace as a result of the mining boom,' the story ran. 'The latest "paper" millionaires are those with big shareholdings in

Samin Ltd, the stock that opened at an incredible $30 on the Exchange yesterday…Other millionaires created this week are the Adelaide couple, Mr and Mrs Reginald Sprigg, who own around 30 000 shares in the unlisted nickel stock Nickel Mines of Australia Ltd. Within a week, these shares jumped from around $50 to about $160 in unofficial trading although they have now fallen slightly.'

Reg and Griselda's newfound status remained virtual since Reg chose not to sell any of his stock, however there were plenty of others anxious to cash in. In November the University of Adelaide had sold 4000 shares in Nickel Mines of Australia at $21 per share for $85 000. It was part of a parcel of 8000 fully paid shares that Reg had donated to his Alma Mater back in 1955 when they were worth just five shillings (50 cents) each.

The university very astutely monitored the boom and sold off another 365 shares in February 1970, when they were fetching almost $80 per share. It was only a couple of weeks after Poseidon stocks had reached their peak price of $280.

After that it was all downhill and many small investors were left licking their wounds and wondering how different things might have been had the Securities Industry Act, which made insider trading illegal, been enacted before all the tumult began. In the summer of 1969/70 there was no law against insider trading and much of the Poseidon boom had been fuelled by an erroneous estimate of nickel reserves—which were ultimately shown to be 1.9% rather than the 3.5% originally announced—and insider trading by company directors and consulting geologists.

Reg chose not to sell any of his suddenly fabulously valuable Nickel Mines of Australia shares because he found the idea morally repugnant. He knew that the share price was ridiculously artificially inflated and the value of the metal in the ground could never match such expectations. He had always abhorred what he called the 'Grenfell Street miner' who exploited geology to play the stock market, and he shunned the very idea of joining their ranks. He also actually believed in the prospects of his Wingellina nickel deposit and hoped that a mine at the site would go ahead, despite the difficulties. Holding on to his Nickel Mines of Australia scrip seemed much the most responsible option.

And besides…he had no immediate need to convert his investment into cash. He and Griselda lived in comfort. There were trips overseas and both his children were pursing their private pilot's licences. There was even enough left over for luxuries like building a new house on the seafront and the lavish but decidedly quirky Christmas present he had bought for his wife at the end of 1967.

Like anyone at the top of their game, there was no inkling, then, that it could all go so terribly wrong.

Spriggina floundersi

CHAPTER 10

AN OUTBACK WILDLIFE SANCTUARY

AT THE DAWN of the 1970s Reg had two ambitious new projects on the table before him. One was a natural extension of Beach Petroleum's expanding oil and gas search overseas. The other was something completely different; a serendipitous entry into an entirely new field. Reg Sprigg—uranium miner and oil explorer; one who surveyed the earth's mineral wealth and calculated how best to exploit it, was turning green.

On the surface it was an enigmatic personal reinvention. But for those who really knew the man it was not at all a surprising turnabout. Reg had always been an advocate of responsible multi-purpose land use. He could see no reason why mining and farming; conservation and cultivation, for example; could not sit comfortably side-by side on the same patch of dirt. As long as each activity was performed with thoughtfulness and consideration for the other stakeholders, then the earth should be used to its fullest potential. And that didn't mean environmental pillage for short-term wealth and gain, it meant being cognizant of the earth as a complete ecosystem and mindful that some areas could be treated more cavalierly than others.

Right back to his student days Reg had had an awareness of the impact of environmental damage—from the 'scorched earth' clearing practices of many early agriculturalists, to the disaster of rabbits on the delicate, drought-prone farmlands at the northern perimeters. It was an awakening exampled by Professor Mawson, who eagerly participated in uranium mining and timber harvesting yet also advocated sustainability. It was Mawson who religiously monitored the growth—or rather lack thereof—of the slow-growing native *Callitris* pine in the arid northern Flinders, and lamented to young Reg how nibbling feral goats and the perennial dry were defeating its best efforts to grow good and strong. It was the professor, too, who insisted they trek all rubbish out with them when they broke camp, despite the scorn of some of his students who couldn't appreciate the effort.

IT HAD BEEN late 1967 when the pastoral lease for Arkaroola Station had come on the market. When Reg heard news of the impending sale he made entreaties to the government to secure the lease and thus protect the area in the national interest. The northern Flinders sheep station contained geologically unique monuments and formations that deserved state government protection, he implored. Arkaroola's rugged hills were home to the threatened yellow-footed rock wallaby he pointed out, and with its legacy of uranium mining at Mount Painter, the country was also rich in social history.

It was in 1966 that Reg had taken the family to Arkaroola for the first time. Smiler Greenwood had accompanied them, and Griselda and the children had loved hearing Smiler's stories of traipsing the Arkaroola hills with his late father, old WB. He told them how, in 1937, he and his brother, Bentley, had convinced the government and the surrounding leaseholders to give them the unwanted no-mans-land that became Arkaroola, in exchange for ridding its rocky ranges of hundreds of feral donkeys, camels, dingoes and goats. And he told them how they had named their new station after the great snake, Arkaroo, of the local Adnyamathanha people.

The Dreamtime serpent, so the story goes, slithered into the northern Flinders Ranges after drinking Lake Frome dry. Its sinuous body carved out the Arkaroola gorge and where it stopped to pass water, so the waterholes—Bolla Bollana and Nooldoonooldoona—were formed. He then retreated to his lair in the Gammon Ranges where the rumblings of his belly still reverberate through the hills.

Margaret and Douglas were entranced. Arkaroola was a place that they had been hearing about all their lives. It was wonderful, finally, to see it for themselves, and to have it narrated for them by one of the station's founders, no less. In an inexplicable way it felt almost like a homecoming. Griselda felt it too. A feeling of belonging in a place she had never visited before.

Once past the station homestead the road petered out and the going became rougher and bumpier as they made their way north into Mount Painter country. Reg pointed out how the undulating hills of the southern Flinders—striped in their uniform sedimentary layers—gave way to the craggy older peaks of the northern Flinders, which had been formed 1600 million years ago and then squeezed from deep underground to now poke up at the surface. It was obvious that these ranges had a special significance for their father as he recalled camping trips with Mawson and his first surveying assignment during the wartime search for uranium.

As they approached Mount Gee, Reg told how it was known as 'the crystal mountain' for its colourful encrustations of purple amethyst, green fluorite and clear white quartz. However, as they drew nearer they were appalled to find that great chunks of the hillside had been blasted away. Rampant rockhounds and mineral collectors had unleashed explosives at the crystal mountain to release sparkling samples for their own pleasure. The evidence of this wanton destruction infuriated Reg. It brought to mind a similar act of environmental vandalism that he had witnessed back in 1944.

When the search for Mount Painter uranium was on and the government's men were camped in these hills, a family of yellow foots had befriended them. The wallabies became so tame that they would take food from an outstretched hand. When the search was called off at the end of the war the men packed up their equipment and left in a convoy of vehicles. Reg saw the wallabies gathered at the side of the track, as if in a line of honour, and then saw guns raised from a vehicle in front and the family annihilated in a volley of masochistic target practice—two of them skinned for their pretty pelts, the others left to rot. It made him feel ill. Senseless bloody destruction of such beautiful creatures, he cried. He had never forgotten it.

Shaking his head now at the rubble at the base of Mount Gee, Smiler, too, was dismayed. But there was nothing he could do. It was almost 20 years since a Greenwood had last been in charge at Arkaroola.

Smiler's brother Bentley had sold the station in 1949 and moved his family to Orroroo where schooling for his two children was available and the rainfall was not so fickle. The following year was a good one and wool prices soared to £1 per pound of wool. The new owner of Arkaroola held the station for a mere eight months—long enough to harvest one wool clip—then sold for a handsome profit. But that was the way of it in this 10-inch-or-nothing country. It was farming not for the faint-of-heart.

In December 1967, when Reg heard that Arkaroola was for sale once again, he was galvanised into action. Now was the time to garner some protection for this fragile native environment. He pestered the state authorities to take action—to no avail. Finally a frustrated public servant declared the obvious, 'Look, the property is for sale,' he told Reg. 'Why don't you buy it and do something yourself?'

Reg knew that it was the only real solution. If he wanted to see Arkaroola preserved he would have to do it himself. And just like with all of his previous grand schemes and ideas, he was willing to stand behind the strength of his convictions. He would invest his own money to secure his latest dream.

And so, on the 18 January 1968, for the sum of $156 000, Reg and Griselda Sprigg became the new owners of pastoral lease number 2240, comprising 610 square kilometres of spinifex-covered hillsides and red, rocky peaks. Reg presented the lease as a Christmas present for Griselda, who had been fantasising about a Jaguar perhaps or an Aston Martin, but nimbly changed gears and accepted the gift with grace. Typical Reg. To make such an erratic decision without consulting with her first. But they had spoken often of Arkaroola in the past few months and he had correctly gauged that she was not averse to the idea of acquiring the country property. She knew, too, how important Arkaroola was to Reg. It was the seat of shared kinship between he and Mawson and had always held a fond place in his heart. And now it was theirs.

Together they brainstormed how best to go forward. They decided that transforming Arkaroola from an ineffective and unprofitable sheep station, to a tightly managed, low-impact tourist facility was the best way for Arkaroola to pay its way. For Reg had taken a mortgage to finance the purchase and they couldn't afford to keep it merely as a token—an outback bauble to hang on Reg's tree of environmental righteousness. Besides, Reg was a businessman and he sensed the potential to make Arkaroola self-sustaining. He envisaged a tourist resort where recreation could go hand-in-hand with environmental conservation. The term 'ecotourism' had not yet been coined and there was no neat and tidy term for what Reg hoped to accomplish. He dreamed of a place where families and students could come to absorb the majesty of the arid ranges. Where they could soak up the grandeur of the ancient hills and learn about the formation of the earth and this unique part of it. A place where the shy yellow-footed rock wallaby could roam in safety and the stately *Callitris* pine take root undisturbed.

It was an ambitious plan. Admirable but far-fetched. Vintage Sprigg.

For a property 640 kilometres from Adelaide—over dirt roads much of the way—it would be an accomplishment even to induce visitors to come. Accommodation and facilities would have to be provided for them, roads graded, water supplies secured, a power station constructed…the list was endless. Arkaroola was currently a run-down sheep station with 2500 sheep, a couple of bores and station tracks and a dilapidated homestead. Turning it into a tourist resort of any kind would be a hard task—a full-time occupation for anyone.

But of course Reg had genuine work projects to devote himself to, and Griselda was tied to the city. Margaret and Douglas were both at secondary

The Sprigg family home, 'Arkaroola' house, on The Esplanade at Somerton Park under construction, circa 1967.

school and the entire family would soon be moving to their new beachfront address.

Almost two years in the making, the brand new $200 000 Sprigg home on the Esplanade at Somerton Park was finally ready for occupation. It was only two kilometres from their Baker Street home but miles away in style and character. Reg had designed it himself—a white, ultra-modern, two-storey, cantilevered, concrete cube with jutting floor-to-ceiling windows overlooking the sea at the front and the swimming pool at the back. He had modelled it after actor Gregory Peck's Mexico City home. To commemorate the completion of this project and the commencement of the next, they named the house 'Arkaroola' and affixed this to the side-street-face of the house in metre-high flowing black letters.

While the family gradually took up residence on The Esplanade, improvements began in earnest at Arkaroola. The first priority was water. Geosurveys staffer Dennis Walter was appointed as manager and Geo-

Arkaroola house on the seafront in 2007 when it sold for more than two million dollars. The downstairs room at right is a later addition to the original structure. *(Kristin Weidenbach)*

surveys's drilling expertise was used to sink five new wells into Arkaroola's hard rock base. The first was dry. The second yielded water too salty, and the third spurted briefly, then failed. At a cost of $15 per metre for drilling, $20 per hour for lining the hole with casing and $100 per day for water testing, just making the 100-metre deep holes was an expensive exercise. And this was only the beginning. The tyranny of distance guaranteed that everything about Arkaroola would be expensive.

The fourth well, at a new site at the station homestead, flowed 5000 litres per hour of semi-brackish, but usable water, and the fifth, too, produced water at the same flow rate suitable for general domestic use. The success of this Copper Creek bore now gave an option for the location of the new Arkaroola village.

Reg had decided from the outset that the tourist motel would not be located at the site of the original homestead. He now selected an area of flat, high ground east of Copper Creek as being the best place for the new development. It was a square-kilometre knoll encircled by a loop of the Wywhyana Creek and flanked on the east and west by spectacular quartzite ridges. To the north one could see the dark hulk of Mount Painter squatting in the distance while to the south the rounded hills formed from old seabed layers loped off towards the Gammon Ranges. His village would serve as a gateway to the unique granite hills to the north—an entrée to the most ancient mountains in the state.

With the site for the village decided, building could begin. Plans were laid out for a bar and dining room, adjoined to a motel reception desk and a small shop cum service station. The first accommodation unit was a row of four transportable cabins that Reg named Thomas Wing, in honour of good friend and fellow Mawson pupil, Richard Grenfell Thomas. They were clean, simple cubicles with two single beds, a wardrobe and a bedside table, and served by a common amenities block.

The biggest building to take shape was a concrete-block powerhouse to house a Ruston and two Blackstone slow-revving diesel engines. The plant had been purchased complete from the township of Penola in the South East when the town was incorporated into the state electricity grid. Reg and Griselda also purchased the services of the plant's caretaker and service mechanic to train 13-year-old Douglas in its operation. He was fascinated with the indefatigable throbbing beasts—as he was with anything mechanical—and begged his father to be allowed to travel to Penola to accompany their relocation to Arkaroola.

Soon there was at least one large truck or semi-trailer making its way the 600 dusty kilometres to Arkaroola each week and the remote outpost was bustling with construction. Fences on neighbouring properties sometimes had to be cut to allow through some of the wider loads, such as sections of the new 20-room accommodation lodge. Once joined together on site the new lodge would form a square resembling a typical Australian homestead, with a wide verandah all around and high louvered windows ventilating an interior atrium. Each room opened onto this central lounge and Reg imagined that here he would show films and deliver lectures on the geological history of the last billion years.

AN OUTBACK WILDLIFE SANCTUARY

In October 1968, Arkaroola was officially opened and Smiler Greenwood snipped the ribbon to open Greenwood Lodge. A lot had been achieved in a short time, but still there was much to do. As tracks throughout the station were upgraded a steady stream of curious visitors came to discover Arkaroola. Their interest heartened Reg and confirmed to him that his instincts had been correct. People were keen to experience and learn about natural history if they were only given the facility and encouragement to do so. If they could appreciate the antiquity and fragility of their surroundings, he believed, they were more likely to respect and cherish it.

Nevertheless, for Reg and Griselda it was a race to keep one step ahead of the tourists and the amenities that they demanded. Throughout 1969, whenever Reg could be spared from Geosurveys and/or Beach commitments the consultant geologist and company director was on the grader at Arkaroola scraping out new tracks, or laying pipes and cables in the expanding caravan park, or installing guttering and huge rainwater tanks to catch every drop of water should it ever choose to fall from the sky. At home in Adelaide, Griselda began to feel that she never got to see her husband.

It was about to get a lot worse.

THE SECOND PROJECT on Reg's plate at the dawn of the 1970s was 9000 kilometres away from Arkaroola and would result in him living away from home for six months in every twelve for the next nine years. Once again it centred on the eternal hunt for oil.

Beach Petroleum's exploration permits already covered 165 000 square kilometres across Australia and New Zealand. Geosurveys/Beach surveying and drilling crews were operating in the Simpson Desert; the St Vincent Basin, encompassing Adelaide's Gulf St Vincent and the Yorke Peninsula; and the Darling Basin in western New South Wales, west of Wilcannia. Farm-out agreements extended Beach's interests into the Gambier-Otway Basin in South Australia's South East; eastern portions of the Canning and Fitzroy Basins in northern Western Australia; the western part of the Amadeus Basin in the Northern Territory, west of Alice Springs; and the southern portion of western Cook Strait, extending onto New Zealand's

South Island. In conjunction with Aquitaine Petroleum, Beach had already drilled an offshore well, approximately 80 kilometres from Nelson using a semi-submersible rig capable of handling the rough seas characteristic of the area. New leases had also just been secured over the Chatham and Pitt Islands, 1200 kilometres east of Christchurch. Gravity and geological surveys were underway and the area looked promising.

Beach's most ambitious new prospect, however, was licence 831 comprising 500 square kilometres in far-off eastern Turkey. The acreage was in the deep Diyarbakir Basin at the foot of the Taurus Mountains and contained several structures where the surface limestone folded into gentle anticlines rising 100 metres or so out of the basalt plains. Immediately to the south of the tenement the giant international Shell company was already producing 2000 barrels of oil each day from similar anticlines on their property. The closest well in Shell's Kurkan Oil Field was only a kilometre from what could be Beach's southern boundary and their Beykan oilfield was only a little further south.

It sounded like a good opportunity for Beach. So in February 1970, Reg and fellow director Hugh Morgan were dispatched to Turkey to consider the deal.

'Griselda, love, this trip with Hugh Morgan will be indelibly inscribed on our memories,' Reg wrote home to his wife from Ankara. 'What a fiasco. Nothing, I repeat, nothing has gone right. Catching aircraft all hours of the day, missing connections, planes cancelled within an hour of departure, luggage lost in Tehran. You name it, we've experienced it. I have lived in my grey shirt since Sunday (it's Thursday evening). Because of the schedule it's had only one dubious wash.

'Tomorrow morning I leave for Diyarbakir at the foot of the Zagros Mountains. Still no sign of my bags. Our interpreter went out buying clothes. I now have an overcoat and new set of (too long) long johns, socks, a green shirt and gold tie. Sounds colourful. I really miss you and the kids. Even a week is too long to be away from the woman I love. I've written some 70 letters since I left so if I've missed anyone it's too bad.'

Despite these travails it was an interesting trip. Reg was thrilled with the sights and sounds of Diyarbakir—the walled city on the banks of the Tigris River that would become their base. He loved the mix of people—Kurds, Turks and Americans—the scenery, the history and the architecture that wove the rich tapestry of the ancient city.

He was also enthralled by the broad-scale geology. In contrast to Australia, which occupied one great continental plate and moved sedately northwards at the rate of 5 to 10 centimetres per year, Turkey sat at the global intersection of two continental plates. As the Arabian plate drove irresistibly northward it collided with Asia Minor, producing devastating earthquakes along the side-slip fault lines and throwing up jagged mountains in the crumple zone where the earth's building blocks crashed together. Right under his feet, in the vicinity of Diyarbakir, the Taurus and Zagros mountains were rising steadily higher as they thrust up and over the Arabian block pushing inexorably from the south.

For Reg it was fabulous stuff. It made Australia seem so passive; so benign, in comparison. He dreamed of having the chance to explore at his leisure.

Fortunately he and Hugh were also excited by the regional geology associated with the leases on offer. Smooth round anticlines formed the foothills adjacent to the city—anticlines concealing thick, oil-reservoir-forming, marine limestones. Reg knew that these structures, with names such as Divana, Hosan and Serbetli were deceptively simple. Often the thick Mardin limestones 2000 metres below the surface were severely crinkled and crushed by the opposing forces of the continental plates. But their potential was great, and the profitable oil-producing Persian anticlines to the south were tapping oil from the very same Mardin limestone beds.

When the Beach Board received a summary of the prospects upon the travellers' return it rapidly gave approval for the Turkish lease purchase and the contract was signed. Within weeks Darby and two Geosurveys geophysicists, Peter Brunt and Anton Bauer, were on site to conduct a gravity survey of the exotic new territory prior to a drilling campaign to begin in May. The team was hopeful that gravity would provide the information they required as seismic or other geophysical techniques were out of the question. Fields were being readied for summer crops across the region and the farmers were restricting access for scientists laying out microphone cables and explosives.

The ploughed up Turkish fields and rough limestone outcrops largely prohibited vehicle access so the Landrover was routinely abandoned. Peter and Anton hauled the gravity meter across country, dodging sheep and the massive, ferocious Anatolyan wolf hounds the Kurdish shepherds employed to tend their flocks. They looked as if they could bring down a bull with ease and made for nerve-wracking field companions. It took about a month to complete the survey, and when Reg joined them about halfway through, he was most pleased with their progress.

'Peter Brunt and Anton Bauer have done a fantastic job with the gravity survey here. They walk 12 to 14 miles a day—surveying as they go—through rough limestone hills,' he wrote home to Griselda. 'The gravity has proved up Divana very beautifully and with my short geological field survey the correlation looks terrific—only hope we get it this time.'

Together they selected the first drill site and then Reg was off to London to hire the drilling rig and organise supplies and equipment.

Back home in Adelaide, Griselda had her hands full placating Geosurveys wives who were unused to having their husbands so far from home. Having their menfolk on short trips to Central Australia or camped along the South Australian coast was one thing, but having them spirited away to the Middle East was quite another. No matter how the men might be enjoying the experience of foreign travel and new customs, the womenfolk wanted them in the familiar folds of home and hearth.

'My dearest darling,' Griselda wrote to Reg in mid-April, 'I spent most of last night on the telephone calming hysterical wives. Poor Heather [Geosurveys secretary] had taken rather a beating yesterday. Mrs Peter Brunt had been on the telephone and torn a large piece out of her hide. She screamed at Heather and shouted abuse. Said all she wanted to know was when Peter would be home. Heather suggested that perhaps if she went out a bit with friends it might help to alleviate the loneliness. She then screamed that she had no friends. (Not altogether surprising). Heather, nearing her wits end, suggested that she would send a cable to Peter from his wife at the company's expense. She showed it to me, quote: "When are you coming home. Barbara." Heather put "love" in front of Barbara.

'Why the hell people feel so sorry for themselves I'll never know. One has to maintain a bright outlook otherwise no one is likely to call or even speak to one. And keep oneself busy. She had not even troubled to write to him. Said as she was on her own she had nothing to write about. I can manage to write reams of piffle to you or anyone else on absolutely nothing if they are away from home. How does she think he feels when no letters come from home?

'Having read of your pleasure at the way the survey had been carried out by Peter Brunt I thought it might be diplomatic to ring his missus and tell her that you were most gratified with his work. I told her it was nice to know that the survey had been completed in record time and her husband would be home earlier, probably well in advance of you. I also said that you would have to return to Turkey about mid-May.

'"But how can you stand it?" she asked. I told her that having married a geologist I had realised from the beginning that if he were going to be a success he would not be able to do his work entirely in an office or his own backyard, and though I did not like being on my own, I had learned to live with this fact uncomplainingly.'

True to her word, Griselda didn't begrudge Reg's time away, but now that she was on her own so much she did find a feeling of aimlessness creeping into her life. With the family now comfortably settled in their seaside home, and with Margaret and Douglas leading busy teenage lives of their own, she began to consider what she might do with herself. She abhorred idleness and was not one to fritter her days away on tea parties and home decor.

True, it was a long time since the days of the Royal Alexandra Infirmary when she had led the life of an independent working girl. But there was no reason why she couldn't work again. Not nursing this time. Sales perhaps. Fashion maybe; clothes had always been one of her weaknesses.

When she mentioned the idea to Reg he responded with an even better suggestion.

'If you want to do sales, why not sell Arkaroola?' he posed. 'We need a city office anyway, somewhere visible to help get the word out. You could run it, and vet prospective staff. It would be perfect.'

Griselda would not earn a salary for her contribution but at least she would be saving the cost of one for someone else. Furthermore, no one would sell Arkaroola's merits with the passion and enthusiasm that a Sprigg would apply. It was the perfect marriage. Suddenly her days went from long and languid to overflowing hours run with a military timetable.

'I don't suppose Reg told you I have started work,' she wrote to her sister in Scotland a couple of weeks later. 'I am doing promotions for Arkaroola and find the job most absorbing and interesting. It is terrific fun. I meet lots of people and it is quite stimulating to be out of the house each day. More interesting than housework. Of course, I still have to do the housework. So far I am managing to cope.

'I get up at 5.30 am, have my swim in the pool, attend to the dogs and groom them, and clean up the lower floor and basement. By the time I get upstairs the family are up and at breakfast. I usually cook Margaret's chop or steak while I have my swim. I set breakfast the night before and prepare the grapefruit and cereal. Then I wash the dishes as I eat my own breakfast. It only means that the family have to be up by 6.30 am instead of 7.00 am. As they are all very enthusiastic about Arkaroola they do not mind a working mum. Before I leave for the office I prepare the dinner and set the automatic oven. Margaret adds the potatoes and vegetables about 5.00 pm and when I get in about 6.00 pm the dinner is ready.'

Arkaroola's new Adelaide office was located in a grand old two-storey mansion on the eastern fringe of the city. More funds flowed out of the Sprigg account to restore Dequetteville House to its former glory, until the limit of their overdraft was reached. Griselda brought furniture from home to equip her office and reception area, and gradually created a style that she was happy with. Geosurveys offshoots Specimen Minerals and Naturelle Jewellery occupied the northern half of the building.

Over in Turkey, by the end of June, Turkish Beach Petroleum's first oil exploration well was underway. Beach's share price rose two dollars in one day on the Australian Stock Exchange, and there was a general feeling of optimism in the company. Geosurveys geologist Bob Laws had been appointed resident well-site geologist and was enjoying the Turkish way-of-life with his Australian girlfriend helping out with company office-work.

Reg, himself, was busy working with Beach's local agent to implement efficient work practices and smooth any troublespots encountered with the local Kurdish population. In a country so different from Australia it was not always easy to do business. Negotiations for rights to operate through Kurdish farm lands were conducted with village chiefs in limestone adobe huts. Sometimes sharing the gastronomic delicacy of sheeps' eyeballs was required to seal the deal. The villagers were friendly and welcoming but chronically poverty-stricken and oppressed. Any opportunities for extortion or thieving were eagerly pounced upon and confrontations with the foreign drill rig operators sometimes turned violent.

'Dealing with Kurds is nerve-wracking in the extreme,' Reg wrote home to Griselda. 'Always there is trouble brewing. You make solemn written agreements and they are looking for ways around—always for more money. It's endless.

'We had a big load of cement arrive at the rig recently. The Kurds decided that they would do the unloading despite protests. Not only are they studiously slow, but managed to burst as many bags as remained intact. Finally the boys had to get the gendarmes in.

'One hidden Kurd in the dark landed a big (12 inch) rock on a rig hand hospitalizing him for 10 days. They throw stones, break rear lights, roll boulders in front of the vehicle and generally menace operations. One mob started fighting over employment and Bob had to get into the melee to break it up.

'Bob and Bill are constantly hunting thieves off from the cement stack. These kids are working their guts out, and worried sick about loose money control by our Turkish reps. The hotel manager was keeping accounts (he is said to be quite unreliable anyway) and now he's disappeared with a go-go singer.'

Reg found that smoothing over skirmishes and continually greasing palms was a frustrating necessity to doing business in Turkey. He even began to suspect that their own Turkish business agent was not entirely trustworthy. Side deals and shady kickbacks seemed to be the order of the day and he began to worry about the huge expenses Turkish Beach was incurring. The nefarious dealings and constant altercations with the Kurds certainly took some of the shine off living and working in Diyarbakir. As the hot, dry summer days rolled on he also found himself unexpectedly suffering from homesickness, and his thoughts turned more and more to the attractions of home.

'Griselda my love, it's the 4 July—a lot of celebrations planned for tonight at the American base, which I will skip. Time is starting to drag now. I am missing you and the kids very genuinely—and only a week gone. I hope I can get someone else to take over in the future. I am quite sick of it, the more so in that my heart for "development" is at Arkaroola. To me it is a wonderful place and I'd like to finish my days close to it. It's something that I hope will keep us together physically and mentally for the rest of our lives.

'Before I left I had thought of getting a few days in Europe or somewhere. At this moment I couldn't be less interested in such touring. To be at home with my family and friends would be the best holiday ever.'

Progress at the Divana drill site was contributing nothing to dispel his gloomy mood:

'The last two days have dragged almost intolerably,' he wrote the following day. 'Things were made a bit depressing by very slow drilling progress since mid-week. Penetration fell from 30 or 40 feet per hour to 5 feet, which is about as slow as drilling can go. Drill bits were wearing out alarmingly so that we were only this morning considering air-freighting a third of a ton of new bits from England.

'It was maddening drilling through a bed of bitumen at 800 feet—plenty of bitumen but it was too viscous to produce. This is the same horizon that produces the really big oil fields further down the gulf!!'

The drill site was 30 kilometres north-west of Turkish Beach's base at the Otel Tatlicilar in Diyarbakir. Radios sent from Geosurveys's Australian base in Woodville for hotel-to-rig communication went astray (like Reg's luggage) and sat in cargo in Rome for 12 days before Geosurveys tracked them down. Meanwhile, dashing from rig to hotel and back again—day or night—was the only way to function. Drilling ground on day after day, seven days a week. With costs of $1000 per day plus rig hire expenses of $3200 per day the company couldn't afford a break. Sleep was taken in short snatches before dawn broke in a searing burst of sun and heat at 3.30 in the morning. The bewildering daylight and the irregular hours of sleep and work induced a state of suspended animation in Reg, and the depressing dirt and poverty all around him threatened to overwhelm his natural energy and optimism. He logged drilling progress in a daze and lived for the next letter from home.

'Did I tell you that two letters arrived from you yesterday,' he wrote to Griselda in mid-July. 'I really needed them. I sat up late last night writing letters, discussing well programs, watching drilling progress etc, and all the time holding back your latest letter for bed. It was the letter I had been waiting for—the Griselda type—full of interest and the travails that cause almost despair but then make up life.

'Sorry I've been so demanding of you re letters. I realised on reading through just what a thorough doing over you're getting during my absence. It gives me enormous pleasure to see you so happy at your work but God, darling, don't overdo it. You're a resilient old cuss (young love) but we all have our limits—I'm at about the end of mine here right now.'

For days the drilling seemed to be advancing agonisingly slowly and always they were just above the horizon of interest; always just beyond it but not quite there. It was excruciating. The rock strata seemed to follow a different order in almost every hole drilled in the region so that correlation with other wells was almost impossible. Just when it would appear that their drill bit was about to penetrate the target oil zone the evidence in the rock would reveal another fold or slippage that pushed the elusive

Mardin limestone further out of reach. Damn and blast these complicated structures.

But finally, by the end of the month, there was some good news from the drill site.

'Griselda love, things are looking a little brighter,' Reg wrote in the afternoon of Wednesday 22 July. 'The promising news is that today we awoke after a sleepless night almost in despair. Well at 6514 feet (1986 metres) in a monotonous argillaceous limestone. We should have been by then in the Mardin reservoir zone, but no.

'We took the risk and decided to core—we still are. The good thing is that white limestone—almost certainly the Mardin, or calcite-filled fractures above it—has appeared. What's more, the fractures are quite heavily oil stained.'

At 10.30 pm he wrote: 'Core in laboratory. *Plenty* of live oil in fractions and right along entire core. Also dead oil and asphalt. Eureka! We're not in the Mardin yet but must be damn close. Now we know at least there's oil here. Bit—$6000 worth of diamonds—in tip-top shape.'

11.50 pm: 'back into hole.'

Early the following morning he elaborated on the previous night's excitement in a second letter to Griselda.

'Honey it looks like this is the day we have been working for since 1954. We appear to have cut good 35-40 (light) gravity oil in a highly fractured zone between 6535 and 6575 feet. It's probably the Mardin, although we felt otherwise at first. We are through the zone and will shortly prepare to test.

'Our decision in the early hours of yesterday morning was almost a desperation move. We had had some dead oil, but nothing live and fluorescent. Waiting to complete coring again tried the patience and the nerves. At 10.30 pm our worst fears were allayed and we could see oil in the core even despite the mud. Once in the laboratory, oil—live and dead—and asphalt showed aplenty. Some of it was nice and drippy. Oh boy!

'Whatever happens in this hole there is now almost a 100% probability that there is producible oil on Divana—what is more the best sites appear still to be drilled.

Barrels of love,
Your Reg'

It seemed that the oil Reg had been chasing for so long was at last to be found in the hills of eastern Turkey. While testing on Divana 1 continued, preparations began to drill a second well at a site called Hosan-2. In between jobs there was a day here or there for relaxing and recreation.

Beach's Turkish business agent, Nihat Boytuzun, owned a yacht that he moored on the Bosporus and used for tourist trips in the Marmara Denizi and the Aegean Sea. It was the thing to do for the upper middle-classes to spend Saturday evenings promenading along the cafes of the main boulevards in Istanbul and its twin city Kadikoy, across the Bosporus, and Reg was invited to join the Boytuzuns in seeing and being seen.

'Being in Istanbul in summer tourist season is something,' Reg wrote home. 'It is a very Mediterranean country and the wealthy disport themselves as ostentatiously as possible—some magnificent yachts, beach houses, clubs etc. Thousands of people sitting and drinking or taking coffee.

'We dined on crays and caviar overlooking the Bosporus. Tasted all sorts of Turkish dishes—eggplant, stuffed capsicum etc. Then onto Nihat's 60-foot yacht, mainly lazing, then by ferry across the Bosporus to the Asian side. It's hard to describe how incredibly beautiful this is. The poplars are green with life, the wisteria is out and all manner of flowering plants. We wound along picturesque country roads and risked our necks on every bend. Nihazt drives the latest Fiat and thinks it is an aircraft. Speed limit is 60 miles per hour, but he sailed along at 105 (180 kilometres per hour), dodging in and out between slower vehicles as though they were dodgem cars. "Don't worry Reg!" is all I ever got to my endless demands to be more careful. Bloody Turks—but I like 'em.

'Motorboating on the sea of Marmara is also unbelievable crazy. We hit one other boat at speed, but didn't sink, so didn't stop!!'

On one occasion Reg was invited to spend the night aboard Nihat's yacht, and later sincerely wished he hadn't. He was awoken in the morning by excited yabbering between Nihat and the fishing crew of a small skiff drawn alongside. The fishermen had somehow dropped their anchor and unsecured chain over the side into the fast-flowing Bosporus edge-waters. Don't worry, Nihat yelled, I have a famous deep-sea diver aboard, he informed them, and summoned Reg forth. He was handed mask, snorkel and flippers and a boat hook to retrieve the chain and anchor, and bundled over the side before he was even fully awake.

The boats were at a depth of only six metres but trying to hook and then haul the heavy chain to the surface on one breath of air was a struggle. On the first attempt, when he did surface, the fishing boat had drifted away in the current and he had to release his hold on the chain to stay afloat. On the third attempt, exhausted from the effort, Reg barely made the surface and took in a great gulp of water as he gasped for air when his head broke clear. The fishermen grabbed the chain and Reg was pulled, coughing and spluttering, back aboard the yacht with much congratulations.

Flying off to London that afternoon and then back home to Adelaide, he barely gave a moment's thought to his Bosporus submersion and what unsanitary horrors could have been floating in the water. Several weeks later, however, when he collapsed at home and was diagnosed with acute hepatitis the conclusion was that he had ingested fecal contamination. Lovely. A long spell in hospital and a prognosis of permanent kidney damage was the result.

'That slowed him down...the hepatitis,' says Bob Laws. 'He was still a vibrant, brilliant man. But before then, he was sort of unstoppable. In my view that did slow him down a bit.'

Thankfully, Bob was still stationed in Turkey and he continued the role of well-site geologist for the Hosan-2 well while Reg recovered in Adelaide.

Bob had fallen in love with Turkey and enjoyed the Kurdish life in Diyarbakir—which was a good thing too, since in typical Geosurveys style, he had been told of his new overseas assignment only days before departure.

'I was sitting on the Chatham Islands thinking, Gee, I wonder what new project Reg has got, and I flew back to Adelaide and within a week I was on the way to Turkey,' he says.

AN OUTBACK WILDLIFE SANCTUARY

Bob had been told that his assignment would be for three months. But like Heli Wopfner had found out before him, Reg Sprigg's three-month jobs often turned into six…and beyond. Bob's girlfriend, Glennys had joined him in Turkey expecting it to be a short-term diversion. Now, after more than six months she was beginning to tire of the male staring and harassment that constantly accompanied the sight of her red hair on the street. More importantly, Bob had come to the conclusion that the oil search in Turkey was a waste of time and money. By studiously poring over the data he became convinced that the area was not prospective.

'The concept was right—it was nestling up close to existing production and had a lot of good things going for it. But I went and looked at the information available in the Turkish Government's mines department—seismic data and other interpretations—and there were some problems,' he remembers.

Reg would hear none of it. He was equally convinced that the oil was there. It was an irreconcilable difference of opinion; one of the reasons that led to Bob leaving the company—on amicable terms—in October 1970.

'Mr Laws's decision to leave our organization after eight years excellent service has come as a genuine disappointment,' Reg wrote in Bob's reference letter. 'Upon reflection, I am conscious that we have asked almost the impossible from him in relation to both Australian and international exploration projects, upon which he has been engaged.'

Perhaps Reg suspected that he had finally demanded too much from Bob, but with so many projects to manage he felt there was no other choice. Beach/Geosurveys was now operating in three countries and Reg couldn't be on the ground in three places at once. In fact, he yearned to spend more time at just one place—Arkaroola—but that, for now, was largely Griselda's domain.

She was flourishing in her role as promotions and marketing officer and had increased her emotional and financial stake in the project by funding most of the village development as well. Long ago, her father had set up extensive share portfolios for both of his daughters and the dividends from his selection of blue-chip investments were now put to good use. Griselda liked to think that her father would have approved.

Increasingly she stamped her style on Arkaroola by decorating the visitor dining room with antique treasures and outback station cast-offs like lanterns, camp ovens and hand shears, which she had collected on her travels and now hung from the beams over diners' heads. She also outfitted the female staff in uniform frocks to her own particular design.

'Style should be A-line (no pleats to get crushed) with round, high neckline,' she advised Dennis. 'There should be a long zip down the back then the dress can be entered feet first to avoid messing up the coiffure. A short sleeve and two front skirt pockets for notebook and pencil, comb or what-have-you. As to colour, the smartest colour choice and popular in various overseas countries was cinnamon. It is flattering to all complexions and never looks grubby.'

With Reg away in Turkey so much of the time, more of the management and administration was now up to her as well. While Dennis was the residential manager at Arkaroola, Griselda was an official company director and the primary Sprigg family point of reference.

Their tourist resort had grown considerably since the official opening in October 1968. New roads to tour the property, an airstrip, a native fauna enclosure and a swimming pool had been added. And in June 1970, their second accommodation lodge was opened by Mawson's widow, Lady Paquita. Regally she stood in front of the low profile, 20-room, brick lodge and named it Mawson Lodge in honour of her late husband. It was a proud day for Reg.

Expenses were still horrendous, in the form of a Cessna 206 for scenic flights and air transfers to Arkaroola, and the endless construction and establishment of services. In addition to the Mawson Lodge, staff accommodation quarters had to be built and a manager's residence. When, or if, Reg or Griselda found time to visit they stayed in a motel room or one of the staff flats, however, Reg had aspirations of building a home for themselves at the foot of a ridge overlooking the village.

AN OUTBACK WILDLIFE SANCTUARY

'This living where nobody else wants to is for the birds,' he told Griselda. 'I have made up my mind to build a nice house for ourselves with a magnificent view.'

Early on he had also entertained other ambitious ideas like establishing an experimental kangaroo farm, setting up a string of overnight hiking huts with bunk-beds and fireplaces across the property, or developing a health spa outpost at the Paralana Hot Springs 20 kilometres north-east of the village on Arkaroola's boundary with Wooltana Station. In his view it could be one of the great inland attractions of the future.

He was not the first to have thought of that. In 1923 a Dr Charles Fenton had set up a primitive therapeutic health spa at the spring. He erected a tent camp at the site where water bubbles naturally from the ground at 62 °C. Sufferers of rheumatism and gout were advised to drink two pints of spring water each day and bathe for 20 minutes followed by 30 minutes complete rest between blankets.

Only one group of patients braved the road trip. No doubt any soothing effects of the mineral waters were quickly erased by the unendurable bumping and shuddering of the 170-kilometre homeward journey over a dirt track in a horse-drawn buggy to the Copley railhead.

In 1946 Premier Playford revived the idea and despatched Reg to accompany the South Australian Minister of Agriculture, Sir George Jenkins and a party of tourism entrepreneurs on a tour of the northern Flinders Ranges from Paralana to the township of Blinman. Bentley Greenwood took some of the officials up to the hot springs in his Buick while Reg drove the remainder in his vehicle. All the officials agreed they had never seen such spectacular scenery before, as that of the Mount Painter country. But the problems of access were obvious and the proposal lapsed again.

For Reg, too, the additional costs involved in setting up a second resort site requiring all the same water, sewerage and power services, made the Paralana idea prohibitive. It was only later that it was discovered that radioactive radon gas also escaped with the carbon dioxide, hydrogen and helium that bubbled from the spring waters. With Mawson, Reg had camped there and swum in the spring in blissful ignorance on many occasions without

235

apparent ill-effect, but a health spa at such a site was unlikely to be popular. It was just as well that the idea was finally abandoned.

An alternative lure for visitors emerged in its place, however. While the Spriggs had been busily turning their arid mountain outpost into a tourist resort, an oil and mineral exploration company, Exoil, had been combing exactly the same ground hunting for uranium. Mining and pastoral leases can legally exist over the same patch of countryside provided that each leaseholder does the right thing and doesn't interfere with the interests of the other. Arkaroola had found the Exoil boys to be harmonious co-inhabitants.

For most of 1968 they were distant neighbours. While Arkaroola village took shape in the southern rangelands, the miners focused their exploration on East Painter gorge. In 1969 the hunt for uranium shifted to the central Mount Painter region, where the World War Two uranium search had first focused. Instead of using the same access track that the government had graded along the creekbeds back in the 1940s, however, which was now washed-out and almost unrecognisable, Exoil elected to cut a new track high into the side of the hills. Bulldozer operator Bill Keen climbed aboard his machine and brazenly carved a sinuous ribbon wending its way along the ridges and peaks, safe from the flash floods that wreaked havoc on the old track.

It was a hair-raising exercise. A steep descent on the track near Mount Painter was scarred by bulldozer slide marks when the Caterpillar D7 slipped over the edge and careered to the valley bottom. Bill hung on tenaciously until the machine slithered to a halt. When he had recovered from the shock he calmly cut a new path out. There were many other temporary 'losses' of bulldozers and under-powered, early-vintage four-wheel-drive vehicles over the steep sides before the 21-kilometre-long track was finally completed.

Reg and Griselda monitored progress and took advantage of the new mountain track to admire the jagged heartland of their 61 000-hectare backyard. They took Margaret and Douglas to the end of the Exoil road, where it plunged frighteningly down toward the Paralana Hot Springs to join up with the company's earlier operations. Sitting on the ground,

AN OUTBACK WILDLIFE SANCTUARY

Sillers Lookout—the pinnacle at the end of Arkaroola's Ridgetop Track.

sheltering from the wind on the leeward side of the jeep and admiring the scene before them, a precipice slightly to the north caught Griselda's eye.

'I wonder why Exoil didn't go up there?' she pondered, half-joking.

It was quite obvious really. The slope from here to there rapidly accelerated to a tiny pinnacle and then fell off into a void over the eastern plains below. She should have known that her impulsive husband would interpret her comment as a challenge.

'Come on kids!' he bellowed. 'We're going mountain climbing.'

The jeep skittered and hopped sideways in the crazy dash for the top. The ground fell away alarmingly on three sides and through the front windscreen was only blue sky. One could only hope there was room enough at the top to stop safely, perhaps even to turn around. At one moment the Jeep seemed to lose all traction and Reg ordered everyone to bail out and shove rocks behind the tyres. Griselda, Margaret and Douglas literally pushed the vehicle to the summit.

The view before them stole the last of their remaining breath. It was like standing on the lip of the world. Far below, the plains in front of them spooled away to Lake Frome, ending in a stark white line on the horizon. At their feet, red, spinifex-covered rock slopes descended like staircases right and left. Further to their left, the purple walls of the Freeling Heights and Mawson Plateau created an enticing north-western backdrop. To the right, a unique view of Mount Painter—its usually hidden northern face. The complete panorama was truly awe-inspiring.

Reg was already considering how Arkaroola could capitalise on such scenery. Back at the village a deal was soon struck with Exoil. In return for contributing to road maintenance costs, Arkaroola would be allowed to conduct tourist four-wheel drive tours along the new ridgetop track. Bill Keen would whip his dozer up the pinnacle and the track would officially terminate at the dizzying peak at the end, where guests would take morning tea and drink in the view. Reg named it Siller's Lookout, in honour of his friend CW (Bill) Siller, Exoil's Chairman of Directors. Thus the Ridgetop Tour became Arkaroola's signature tourist attraction.

Tribrachidium heraldicum

CHAPTER 11

BAD TIMES

It was a harsh old country. There had been good rains in 1968, the first year of Arkaroola Sanctuary's existence, but since then it had been dry. Two inches per year was not enough to replenish the waterholes or stimulate the wildlife. Grandiose statements about native plant and animal regeneration would come to nought if the fauna failed to reproduce under drought-induced stress and the flora lacked the moisture to germinate.

Then…in March 1971…it began to rain. Ten inches fell in three days. It was more than Arkaroola's total annual average rainfall. Forty people were trapped in the village by the loop of the Wywhyana Creek.

In the summer the rain continued, culminating in another dumping in January 1972, when six inches fell within a week. Eighty-three staff members, motel guests and campers were entertained with indoor games and scavenger hunts until roads and airstrip were reopened.

It was an interesting time to be at Arkaroola. Small waterfalls cascaded from Griselda Hill in front of the motel and dusty acacia and eremophila leaves sparkled with a bright green varnish. When the first fat drops plopped onto the ground at the village the motel staff gathered at windows and doorways completely unaccustomed to the sight. They took beers outside and cavorted in the rain wondering if this was really *it!*—the breaking of the drought. Guests looked on—bemused—and worried about their onward travel plans.

The rain brought new life, but new problems, too. Feral goat numbers exploded out of all proportion threatening to overwhelm any native plant or animal regeneration. They competed with the yellow-footed rock wallabies for fodder and nibbled off all the delicate new shoots they could find. Reg invited the sports shooter association to embark on a culling program. The 2000 to 3000 goats they shot each year was barely enough to offset the natural increase in population numbers…But enough to invoke the ire of urban animal liberationists. If only the antagonists would come up here and see for themselves the destruction that the feral pests wrought, Griselda lamented. How else to try and protect the native ecology?

The goat battle was just the first of the many problems that came to define the Sprigg 1970s. Arkaroola's accounting was in a mess. Costs continued to skyrocket. The project had yet to make enough income to cover expenses in any one year. Reg found it impossible to keep track of operations from Turkey and was confronted with bewildering financial figures when he briefly returned home. He needed to continue working to provide funds for Arkaroola, and yet he feared that his time away was potentially placing the business in jeopardy. He needed a manager that he could really trust. Already two had been 'released' for questionable practices.

In March 1973, Clyde Thomson was appointed manager. Clyde had been working for Arkaroola almost since the project's inception and gradually assuming more responsibility along the way. He was a commercial pilot and already in charge of all of Arkaroola's flying operations. Although he was only 28 years old, Reg felt sure that Clyde had the skills to nurture Arkaroola. He had the right mix of outback experience, people skills and level-headedness that could put Arkaroola on a sure footing. Furthermore Clyde was almost part of the family. Reg had given him his first job and trained him as a geological field assistant and surveyor.

'I started with Naturelle Jewellery—polishing gems at Woodville,' Clyde recalls. 'I would've only been about seventeen—straight out of school. We were the first ones to do it. You know, typical Reg: let's do something we don't know about and work out how to do it after it's done. Reg was always a "Fire, Ready, Aim," person. He never really did the implementation plan. If he saw something he wanted to do, he'd say, "well, we'll start doing it, and we'll learn how to do it on the way." He had a circle of people around him and 99% of the time they would get it right in the end—just through his determination.

'I then went to work for Geos themselves,' Clyde continues. 'I didn't have enough money to go to university. I had to do all my surveying at night school and at the camp—sitting down with the hurricane lamp doing my reading, while I was a field assistant to Bob Laws. We were down at Millicent in the South East drilling for oil. We were both on our first jobs. He would've only been 21 or 22.

'Then they transferred me to surveying. And then I became a driller as well. When the opportunity came to do the airborne survey I asked to do that because eventually I wanted to be a pilot. The first job was to go and work with Helicopter Utilities with Allan Vial—"Vile Allan" we used to nick-name him,' Clyde laughs.

Reg's old friend from the South Australian Department of Lands aerial mapping unit was now working for a company called Helicopter Utilities, which had been contracted by the Commonwealth Bureau of Mineral Resources to conduct a gravity survey over Cape York Peninsula. Allan's company would provide the vehicles, aircraft and pilots. Geosurveys would provide the gravity readers, plotting equipment and geophysical personnel.

Since young Clyde had aspirations of becoming a pilot he was keen to be selected for the airborne gravity survey. Reg and Darby readily agreed. Through serving unofficial apprenticeships in all of Geosurveys's divisions Clyde was gradually being trained as a junior operations manager. By 1966,

Geosurveys gravimeter operator Harry Reith is deposited via helicopter to take a gravity measurement. Geosurveys employee Clyde Thomson was conducting similar work when his helicopter crashed on Cape York Peninsula in 1966.

after four years with the organisation he was already an experienced gravity operator and surveyor, and he would be a valuable addition to the air-survey team.

Two helicopters, two planes and two vehicles were involved in the survey. The vehicles took gravity measurements along the roads. The helicopters took gravity measurements in remote and inaccessible bushland, and the planes did some gravity surveying as well as flying in fuel and supplies for the camp.

Each day certain sections of bush would be covered in a grid pattern. At the central point someone with a barometer would take atmospheric and density readings to tie all the measurements together. The helicopters would then fan out from this point to eight different gravity sites. At each site the helicopter pilot would bring his machine down in a small clearing and Clyde or his equivalent, Harry Reith, would jump out with his gravity meter, allow a few minutes for it to settle, take the gravity reading, and climb back on board. The whole procedure taking only 10 or 15 minutes. It was exactly the same routine as Jonas Radus was using down south, sheathed in the Geosurveys diving chamber under the waters of Gulf St Vincent.

For safety monitoring, at each landing point the helicopter pilot would radio his position to a central call centre. It was also routine procedure that the pilot carry an axe. Sometimes the landing points were barely big enough for a helicopter to drop into vertically. Taking off required more room and then the pilot would have to clear scrub and overhanging branches to fashion a small airstrip.

On the afternoon of 4 June 1966, Clyde and his pilot, Jim, refuelled and took off to start a new grid pattern. It was far too wet to land at their first survey point so they changed direction and headed for a new site. Jim radioed their change of plans to the call centre but apparently the message didn't get through.

On the ground, Clyde took his gravity reading and climbed back on board. It was a small clearing they had landed in but Jim gauged it to be big enough. He was an experienced pilot; ex-Navy. Everyone agreed he was good. On this occasion he misjudged.

'The clearing was too small and the trees were too big to chop down,' Clyde remembers. 'He thought he was going to climb out. But we had trouble with the engine. We didn't have enough power, and the main rotor blade hit the top of a tree.

'The helicopter hit the ground, bounced and rolled over. The petrol tank behind me exploded. Petrol went over the engine and then there was an explosion which blew me through the canopy as though I was in an ejector seat. The pilot was trapped so I went back in to help him out and the petrol tank behind *him* exploded, showering us with burning petrol.'

Clyde dragged the pilot free of the inferno. They were both badly burned. Clyde had first- to third-degree burns to 40% of his body. His long trousers had partly protected his legs but his nylon shirt had melted to his body and his face and head were aflame. The pilot was worse. He had just removed his shirt and was wearing only a pair of shorts. When Clyde went back in to get him he was slumped in his seat, on fire from head to foot.

After dragging him a safe distance away, Clyde gathered the scattered contents of a first aid kit that had been thrown clear and attempted to tend the man's burns. He found a mangled tin with which to fetch water from a creek a quarter of a mile away.

'We crashed about four o'clock in the afternoon. During the night I kept him alive. I was using him as a reference to get water out of the creek. I had a little peach tin that exploded, and I'd get water out of the creek, and he'd call out to me so I could find my way back. And on the last trip he didn't call out. I managed to find my way back. But he...he'd died.'

Clyde was on his own, injured and traumatised, in the middle of remote tropical bushland. It was another 12 hours before help arrived.

'In the morning, I thought, they'll come and find us. But they didn't, because we'd changed our position. They kept flying over. About four o'clock again, Harry Reith, flying with his pilot, Harvey, said, we'll go back another way. Harvey said, no, we're running out of light. But Harry persisted. And just as they went past a bit of sun shone on the perspex and they saw us.'

Clyde was picked up in a helicopter and taken to Aurukun Mission, 50 kilometres away, on the north-west coast of Cape York. From there, a DC3 pilot—taking off illegally at night on an airstrip illuminated with car headlights—took him down to Cairns Hospital, where they were accustomed to treating burns suffered by the sugar canecutters. From there he went to Brisbane hospital for a while, and finally to Adelaide.

And it was there that Reg and Darby blustered into his room.

'Well, what do you want to do now?' they asked him. 'You're burnt, you probably can't survey.'

'Well, what I want to do is fly,' Clyde replied.

'Right, that's fine,' Reg said. 'We'll fix it.'

'Any other organisation would've probably said, well that's okay, when you leave you can do that,' says Clyde, 'but Geosurveys, with Reg, it was like a family.

'The first thing they did when I got out of hospital, they took me to the airport and put me in the company plane and said, well here you are, have a fly. I had had lessons before. I guess that was a test of whether I really wanted to do it or not. I said, fine. So they made a commitment. I went off and did a 12-month course and did all my theory subjects, and Geosurveys paid me to do that.

'I think that says a lot about Reg. They helped me to get it in perspective. They'd say, righto, this has happened, now come on, we wouldn't make this the major obstacle in our lives, so dry your eyes and get on with it. And that was the attitude.

'I was in and out of hospital for that 12 months, having grafts to separate the webbing on my fingers and so forth, but in between hospital I went to theory classes for aviation at Parafield.

'It was something about them—Reg and Darby—looking after you. It was more than just an organisation, it was a family in that way. They were looking after the individual. It was a personal relationship rather than an organisational relationship.

'Reg Sprigg was the father figure of the company. Reg was the patriarch of the organisation and Darby was the mentor that made sure everybody could develop to their fullest extent. Both of them were tough, don't worry. If you did something wrong or foolhardy then there was no mercy. But if you made an honest mistake, that was fine. They'd pick you up, dust you off—or wait for you to get up and dust yourself off—and get you going again.'

For his ordeal Clyde ended up with a pilot's job at Geosurveys and the Imperial George Medal, the second highest gallantry award that a civilian could receive. He is one of only 49 Australian civilians to have received the medal since it was instituted by King George VI in September 1940 at the height of the Blitz.

'Mr Thomson displayed great courage and bravery and complete disregard to the safety of his own life in going to the aid of the pilot and attempting to save his life,' the citation reads.

He is now Executive Director of the Royal Flying Doctor Service in Broken Hill.

When Clyde took over as manager of Arkaroola, back in early 1973, he already knew that the project was in trouble. For some time he had been telling Reg that this one or that one was not doing the right thing by the organisation. During the two-and-a-half-hour flights to or from Arkaroola, he would warn his boss of cost overruns and financial haemorrhages. But Reg had problems pressing in on him from all sides.

Small amounts of oil had been found in each of the seven wells drilled in Turkey so far, but none had proved to be the commercial discovery Reg had hoped for. Four million dollars had been poured into the Diyarbakir foothills with not a single producing well to show for it.

Back home things were equally moribund. The Commonwealth Government Petroleum Search Subsidy, which had been providing up to 50% subsidy for drilling and exploration costs, had been reduced to 30% following the exciting oil and gas discoveries of the early 1960s, and virtually repealed completely in 1969. The entire industry had slumped into the doldrums. Low prevailing prices and other punitive policies of the Whitlam Labour Government provided little incentive for oil companies to explore. Only 11 wells would be drilled in South Australia for the whole of 1973, compared to 29 the previous year. And things were only going to get worse.

For Beach it was time for cost cutting and consolidation. The company had bought Geosurveys and all its operations and now decided to shed unprofitable divisions. Plans were made to sell *Saori* and dissolve all marine activities. The Woodville workshops would be closed and excess land disposed of. Buildings, vehicles and equipment would also be sold to boost Beach's coffers. Thirteen of the 50 Geosurveys staff had been let go and at least five more retrenchments were planned.

Beach was left with a skeleton geological staff providing one geologist per continent and one fully engaged in outside contracts. A geophysicist, a driller and survey staff employed on outside contracts were also retained. It was a painful exercise for Reg, watching the organisation he had built from a one-man dream, being dismantled in such a cold commercial manner.

Even Reg, himself, was not spared. Somewhat at his own suggestion he had stepped aside and been replaced as General Manager of Beach Petroleum. Some time ago he had felt that a specialist petroleum engineer might better serve the company's evolving requirements. Reg's talent had always been for geological fieldwork and now he felt that someone with practical experience in oil and gas field development could inject new vitality into the company's exploration efforts. The Board agreed. However, the replacement had turned out to be a disaster. In October 1972, Reg resumed responsibility but by then everyone felt insecure, uneasy and resentful. Company morale was depressingly low.

Life limped along.

At Arkaroola, the once-longed-for rain continued to fall. The goats multiplied, the roads washed away and the tourists stayed home.

By the end of 1973 the position was dire. Arkaroola mortgage repayments were in arrears, suppliers were refusing credit. Reg appealed to the government for assistance but none was forthcoming. He couldn't even sell some of his nickel stock to bail Arkaroola out. The nickel boom was long over. His Nickel Mines of Australia shares, valued at $160 each in January 1970, were worth just $1.75 at last count. He and Griselda had poured $900 000 cash into Arkaroola and were trying to service a $400 000 overdraft. They had nothing more to give.

At Arkaroola, Clyde studied the accounts and presented his recommendations to Reg.

'You were always straight with Reg, you wouldn't sugar coat. I said, "Reg this is the way it's going and it's not looking good." And he'd say, "Well, show me, show me". He was very impatient.'

Clyde told him that building Mawson Lodge had soaked up all available money, and maintaining Adelaide business operations in the stately Arkaroola House on Dequetteville Terrace was eating up 25% of income.

'Reg, it's time,' Clyde told him. 'It really is time to figure out what you want to do with the place.'

The first decision Reg made was to sell Arkaroola House.

'It was terrible,' Clyde recalls now. 'It was a lovely, old building. People would go there and be very impressed, and it very much appealed to Griselda's sense of style. It was her. She ran it and everything else, and it was her image. He went to Griselda and said, "Look, here's the figures from Clyde. We're going to close it." And she never forgave me for that.'

But selling Arkaroola House was not going to be enough to extricate Arkaroola from the financial quicksand. More drastic action was required if all debts were to be repaid.

Reg assembled the family together and laid the situation out in black and white. He and Griselda regarded Arkaroola as a family project and felt that 21-year-old Margaret and 19-year-old Douglas should have a say in its future, too. They were both working for the sanctuary—Margaret in the Adelaide office and Douglas at Arkaroola, running the powerhouse, driving tours and making the weekly supply run to the city—and Reg hoped that his children would both devote their careers to his self-styled brand of environmental tourism.

There were two options, the patriarch announced: Sell Arkaroola—for which there was general disagreement; or keep Arkaroola and sell everything else. And he really meant...everything else. Arkaroola House, the Cessna 206 aeroplane, their home on the seafront, the land at Coobowie on the Yorke Peninsula...all their assets would be gone. Arkaroola Sanctuary would remain as their one family treasure.

The decision was unanimous.

Although they were united in their desire to keep Arkaroola, the fire-sale invoked different sorrows in each of them. Naturally, Griselda was forlorn at the prospect of losing Arkaroola House, but much as she almost hated to admit it, she was also sorry to leave their home on the seafront. Yes, she had complained about the endless housekeeping demanded of the place, but it was a fashionable seaside address, and they had hosted some marvellous parties there. Moving the family to a rented two-bedroom flat would put a lid on entertaining for a while.

For Douglas, it was the plane. He felt it keenly. Aeroplanes were at the centre of his being. His earliest memory was that of sitting in the Anson and burning his two-year-old bottom on a piece of metal between the seats—most probably freezing cold steel after the plane had returned from a high-altitude mapping flight at 20 000 feet. He had gained his student pilot's licence the day he turned 16 and his full unrestricted licence at age 18. His greatest ambition was to gain his commercial wings and become chief pilot for Arkaroola. Doubling up in a two-bedroom flat was not a bother for him, but seeing the Cessna 206 disappear was gut-wrenching.

Reg worried about his staff. Retrenching people from Geosurveys had been depressing enough. Now he was faced again with the same human winnowing exercise. But there was no alternative if Arkaroola were to survive.

'It is not without sadness that Griselda and I have decided to sell Arkaroola House and our home at 30 Esplanade to assist the continuing financing of the Arkaroola project,' he wrote in a memo to all Arkaroola House staff in April 1974.

'To us the Arkaroola environment and tourist project is the most important development in our experience. We believe it to be of monumental significance to Flinders Ranges tourism, but more importantly to the preservation of what is undoubtedly South Australia's most outstanding geological monument and arid range environmental resource.

'We sincerely believe our decision to be the correct one, and an indication of our determination to assure Arkaroola's success as one of the country's most outstanding environmental projects and places of public interest. We wish all of you who have stood with us to share in the overall satisfaction of achievement in a national cause.'

While preparations for the big sell-off got underway, the rain kept falling. It seemed a confirmation of their approach…that there was no way they could trade their way out of the financial hole, especially when tourists couldn't even make it to their motel with roads closed all across the outback. Reducing overheads was the only way forward.

In the first four months of 1974, Arkaroola received almost 30 inches (760 millimetres) of rain—an amount that would normally take four years to fall. By the end of the year they had had a metre of rain. The Wywhyana

Creek, usually a dry, pebbly walking trail, ran at a depth of two and a half metres for much of the year.

Further to the north-east at Santos's new Cooper Basin gas treatment plant at Moomba, the private corporate village had its worst floods in its brief history. A decade's worth of rain fell and combined with floodwaters swirling down from the north to close all road access for six months. It was the year the outback became one giant muddy lake.

For Reg—ever the optimist—the rain was a Godsend. Despite the family's personal travails there was always something to be thankful for.

'There is one great consolation,' he wrote to the Commissioner of Highways in a letter describing the extensive Easter road damage. 'The effect of these April rains will last throughout the winter into spring. July to November should see the best wildflower season ever. Regeneration of the depleted natural flora and fauna has far exceeded our wildest hopes. The *Callitris* pines are regenerating for the first time in living memory in this area, also many species of previously dwindling eucalyptus, and the marsupial population is proliferating remarkably. In this latter respect we feel that our efforts to regenerate this magnificent Mount Painter country have now been amply justified.'

BY THE END of 1974 Arkaroola's outstanding debts had been paid and the Spriggs were ready to start life afresh from their rented flat at Glenelg. However there were still some tough times ahead; more personal and professional obstacles to overcome.

The year 1975 saw management of Beach Petroleum relocate to Melbourne, *Saori* wrecked, and the termination of Reg's north-west nickel venture.

Almost one million tons of laterite nickel in grades averaging 1.2% nickel, had been proved up at the Wingellina site. In the last five years, Reg's company, Nickel Mines of Australia, had been largely working alone trying to develop cheaper extraction methods for the lateritic nickel. Their business partner, the giant Canadian International Nickel Company, was happy to subsidise their junior efforts for a while, but when world prices and international demand for nickel fell, their indulgence was withdrawn. The partnership was terminated and Reg's 20-year association with the far north-west of South Australia was over.

It was an operation he looked back on with nostalgia—a project that had spanned his evolution from public servant geologist to multi-company director. His two decades of intermittently trudging through the inhospitable rock country that he had named the Giles Complex, after the explorer Ernest Giles who had trekked these ranges in the 1870s, had been a time of simple pleasures for Reg. He was sorry not to have found the more lucrative sulphide nickels that Inco desired, but he wasn't sorry to have tried.

In the middle of the year, Reg learned from the local paper that *Saori* had been lost. She had been bought from Beach by a shark fisherman but the last news Reg had had of her was plans by a four-man Adelaide syndicate to send her into the Pacific to protest against the proposed French nuclear tests at Mururoa Atoll. The head of the syndicate had asked the State Government to help pay chartering fees and expenses as a way of showing South Australian support for the protest.

Reg had heard nothing since, until the announcement in *The News* that *Saori* had been wrecked on the rocks of Wedge Island, 60 kilometres southeast of Port Lincoln, at the mouth of the Spencer Gulf. The skipper and his three-man crew were safe but *Saori* was said to be resting on jagged reefs. With continued bad weather, the experts were predicting that she would soon be, 'nothing but matchwood.' Apparently a fuel line had burst while *Saori* was working in close to the island catching shark bait.

'The fuel line breakage caused complete engine failure and within minutes the stern swung around and we were on the rocks,' said the skipper.

What a sad end for such a fine boat, Reg thought. That she would break up on rocks just south of Spencer Gulf where he had spent so many happy days with her. To make matters worse, rumours began circulating that it may have been an insurance job. But when, a couple of years later, Reg had the opportunity to see the wreck for himself, he was flabbergasted. All thoughts of skullduggery were abandoned. There was simply no way a human being could have orchestrated such a stranding. The hand of Mother Nature had cupped the little ship, lifted her from the water and placed her gently on a rock ledge, warm and dry, high above the reach of the waves. She sat there in perfect dry dock like a model ship in an invisible bottle. It was remarkable.

Staring down on her from the comfort of Arkaroola's brand new Cessna 207 Reg and Griselda marvelled at the sight of *Saori*, standing perfectly on her keel and preserved in seeming miniature, dwarfed in a shallow cave at the base of 200-metre overhanging cliffs. There was no way to get to her, no way to salvage her. She would sit there until she simply weathered away into the sea. Reg was pleased for her; glad that she had not been sacrificed on the rocks for money. This was a much more dignified descent into old age.

MUCH MORE SORROWFUL than the loss of *Saori* was the death of Reg's father, 'Pop.'

Since 1968, when Reg had taken over Arkaroola, Pop had spent many happy days there, fossicking for rocks and hunting for rabbits. He died doing what he loved—hiking the rugged hills with a swag of traps, and a carry bag for his catch. After a day of good hunting he returned to the motel, ate a hearty meal, walked outside, and dropped dead from a heart attack. It was an okay way to go for a man of eighty-two.

Now it was just Reg and his sister, Connie Bagshaw, whom he saw only rarely. Their relationship was strained—the depths of it reaching all the way back to the years of the depression, when teenage Connie had been forced to abandon her private school education when Pop lost his job and money was tight. Reg, three years her junior, had made it all the way through university and Connie resented his success. Their younger brother, D'arcy, had died many years ago, and mother had followed a year later. Now father was gone, too.

But Pop had lived a full life and Reg knew that he had achieved contentment.

'Everybody that sees me says how well I look,' Pop had written in a letter to Reg in July 1970. 'Some just can't understand it after my illness but I am doing my best to look after myself and taking plenty of vitamins and things that my body needs, and have plenty of holidays, as I don't want to die yet as I am still enjoying life.'

And then there was a warning for his son:

'Reg I do hope you will soon take things easier as you are getting to an age where you will have to take more care with your body and be careful about your heart and blood pressure. Try and put it on to some younger person. I know I was going the same way in business at Stansbury, working seven days a week year in and year out. I would not listen to Mum. She could see I was slipping back, but as we had worked up a good business I did not like selling it. So one day she put an ad in the paper, and people came over and said they would have it—just what they wanted. They did not think twice about it. It was a good thing. I am sure I would have died many years ago, and I am still sparking.

Cheerio and keep well,

from Dad'

It was not long before his father's words would come back to haunt him.

By October 1978, Reg himself, was admitting he needed a holiday. The past few years had been gruelling. The continual back and forth to Turkey; the intellectual strain of the hunt for oil; the tumult of near-bankruptcy; the inexorable accumulation of petty problems that relentlessly stalked Arkaroola—it was all slowly grinding him down. He was a man of exuberance and energy, but at age 59 he was gradually discovering—to his dismay—that his stores of energy were not limitless after all. He needed time to revitalise. He needed a break.

Griselda was delighted. 'For the first time in 27 years,' she wrote to her sister in Scotland, 'Reg said that he felt he needed a holiday.'

Together they worked out the itinerary for a cruise of the Caribbean and a trip to America in early 1979.

'This is going to cost a packet,' Griselda remarked to Reg when she had finished typing up the schedule.

'There are no pockets in shrouds,' he replied. 'We might both be dead before January, then it won't cost anything.'

She scolded him for even joking about such a thing. Her father had dropped dead at a young age and she hated the thought of tempting fate again.

It was the perfect time for a holiday. Their lives at last seemed to have reached a rare plateau of stability. Arkaroola had just celebrated its tenth birthday and was almost paying its own way. Winter rains had deterred visitors so that the month of June had seen the worst occupancy figures on record, but at least they were no longer in debt. Management of Beach Petroleum had moved from Adelaide to Melbourne in November 1975 and Reg had been appointed Technical Director on a permanent consulting basis. Although it meant he spent two weeks of every month in Melbourne, at least the trips to Turkey were in the past.

Early in 1977, after seven years of effort, Beach had decided to relinquish its Turkish oil licences. Neither the seven wells around Diyarbakir nor the two years' worth of exploration in the south-east of Turkey, near the border with Syria, had yielded oil in commercial quantities. Bob Laws had been right. There was no oil on Beach's leases.

'We were a few years too late,' Reg conceded, thinking of the hundreds of metres of solid crude they had intersected with their drills—the bitumous oil that had long ago ceased to flow.

'How late?' asked a naive Beach accountant, wondering if they had just missed out on a good opportunity.

'About a million years!' Reg laughed.

Instead of stretching resources on expensive overseas operations Beach decided to refocus on its traditional Australian territories: the Otway Basin and the Simpson Desert.

In 1978 the company was preparing to drill at a site called Colson-1. Reg planned to be there when the well was spudded. Griselda decided to come too. It would be a chance for the two of them to steal some time away together. They could camp out in the Simpson, as they had once loved to do, and make a mini-holiday out of it. A prelude to the real vacation in the New Year.

BAD TIMES

The plan was to drive from Arkaroola to Oodnadatta on 17 October. From there they would drive north-easterly into the desert to the Colson site, between Andado station and Geosurveys Hill, where they would stay for a couple of nights. They would then proceed up the Hale River to Alice Springs and fly back to Adelaide on 23 October. They would follow their Simpson sortie with a few days in Perth, where Reg would attend the annual Australian Petroleum Exploration Association conference and Griselda would do tourism promotions for Arkaroola, and finish up with a couple of days in Melbourne. Altogether they would have 18 days away. It was a long time since they had enjoyed some time together, away from Arkaroola. As Reg said, 'others have private time, why can't we?'

While Reg attended to business in Melbourne and Canberra prior to their departure, Griselda organised their camping equipment at Arkaroola, in between working in the bar and office and turning into a cleaner or receptionist when necessary. She was spending more and more of her time at Arkaroola. Wages and penalty rates were so high that donating her own labour seemed to be the best way to try and keep costs down. Sometimes she complained of the 15-hour days she often put in, but only in jest. She knew that the rest of her family were working similar hours and for all of them the emotional rewards of Arkaroola were still enough to outweigh the negatives.

Still, there were times when it was a hard slog. Quite often she wouldn't see Reg for a month or more, while she was at Arkaroola and he was in Adelaide or interstate. At Arkaroola the rare hours she wasn't working, Griselda spent on her own. She knew that one could not effectively manage staff with whom one also socialised, and the isolation of their little tourist village bred gossip and petty politics that she was best apart from. It was lonely at times.

On the 14 October, three days before they were due to depart for the Simpson, Reg arrived at Arkaroola. For Griselda, it was lovely to have her husband at her side again. They worked feverishly to answer correspondence and clear the decks before their trip, eagerly anticipating the luxury of some uninterrupted time together. But the night before they were due to leave,

Reg received word that heavy rains had delayed the transfer of the drill rig from Moomba. The spudding-in at Colson was on hold indefinitely and thus so was their Simpson sojourn.

On the spur of the moment, Reg and Griselda decided they would have a reprieve in Adelaide instead. Doug was flying down on Thursday for the aircraft's scheduled 100-hourly maintenance service. They could go with him and secret themselves at the flat until they were due to leave for Perth. They could have a short city holiday instead of an outback holiday. No matter, the goal was purely to spend time together.

Over the weekend they demolished the piles of paperwork in their respective 'too-hard baskets' and took breaks sunning themselves on the balcony overlooking the sea. The spring weather was glorious.

Monday, Griselda continued pecking away at the typewriter while Reg met with an officer from the Commonwealth government National Energy Research Development and Demonstration Council (NERDDC). Reg was a member of the Fossil Fuels Section of the council. He left in the afternoon to take the officer to the airport, telling Griselda he would see her at home for dinner.

By seven o'clock in the evening he had not arrived home. At twenty past seven he staggered in and fell straight into bed and began vomiting. He had been struck by a blinding headache on the way home, he said, and pulled over before he blacked out. He couldn't tell her how long he'd been there. She found out later that it had taken him two hours to drive the last 5 kilometres down Anzac Highway. He had pulled into the kerb several times. How he managed to drive home was a complete mystery.

Griselda knew right away it was serious. She called their local doctor, and he called a specialist, and by midnight Reg was in an ambulance on his way to the Royal Adelaide Hospital with a suspected cerebral aneurism.

It was Monday 23 October. If they had adhered to their original holiday plans they would have been en route from the Colson rig site to Alice Springs—somewhere along the creekbed of the Hale River; on their own in the middle of the Simpson Desert, one of the most remote and inaccessible places in Australia. Griselda shuddered to think of it.

BAD TIMES

By Tuesday morning, after a brain scan and a lumbar puncture, the diagnosis was confirmed—a haemorrhage at the back of the skull on the right side, caused by a 'bubble' on the main connecting artery from the cerebrum to the cerebellum. It was a bleed from the brain that could easily have been fatal. For sure, Reg was not out of the woods yet, but Griselda liked to think that their last minute decision to cancel the Simpson Desert trip was a good omen for Reg's recovery. Even if they had been at Arkaroola, the distance from medical help could have been deadly. Instead, Reg was only 12 kilometres from the state's best trauma hospital when the emergency occurred, and he was now being attended by reputedly the most revered neurosurgeon in the Southern Hemisphere. She felt they were blessed.

Indeed, Reg's demeanor that afternoon, less than 24 hours after the stroke, was remarkable. His speech and movement all appeared normal and he was in less pain. Griselda would even describe him as 'chirpy'—flattering the lovely young sister and teasing the nurses. An angiogram was scheduled for the next day, however, and she knew, from 13 years experience as a radiographer and radiotherapist, that the procedure was demanding.

'I can well remember this examination takes a lot out of the patient so I do not anticipate he will be as good tomorrow night,' she wrote to her sister.

To keep her mind occupied and stave off worry, Griselda resorted to what had become second nature to her—writing and answering letters.

'It has been a very long week but not exactly a holiday,' she wrote. 'It seems like a year.

'After [the neurosurgeon] had examined Reg he came out to the corridor to talk to me. He told me Reg will require to be in hospital for at least a month and if brain surgery is deemed necessary or advisable, possibly up to three months. He also said he would have to "change his lifestyle". This is abundantly obvious—even a dynamo can burn out. Reg is just so lucky he has been given a second chance. This is a "warning". Not too many people get one with a cerebral haemorrhage.

'In two ways he has been very lucky—at the time this happened if we had gone to the Simpson Desert we would have been somewhere on the

Hale River. I was interested to know if, in these more relaxed circumstances, it would have happened at the same time. The doctor assured me it would have. I shudder to think of the results. Indeed I know what they would have been. On that Monday there was two and half inches of rain in that area. I had had sand tyres put on the vehicle so we would have been bogged down and by now the crows would have been well fed and our bones would have been well bleached.'

On Monday, a week after the stroke, Reg was scheduled for neurosurgery to repair the original defect and prevent a secondary bleed. He was very depressed, the night before, at the thought of the procedure in front of him. It was quite unlike him. He told Griselda he thought he was going to die.

'We have been courting death and disaster for nearly 30 years and getting away with it,' she snapped at him. 'I don't see why Monday should be any different.'

She was not one for mollycoddling, and normally neither was Reg. But her abruptness masked her own deep concern. The day of the surgery she labelled as 'the longest day of my life.' She felt herself aging by the second as the hours dragged by and she paced the confines of the flat.

The operation was more complicated than expected and took five and a half hours instead of the two to three hours anticipated, but late in the afternoon the surgeon called to announced it had been 'one hundred percent successful.' In the evening Griselda was allowed a 30-second visit. Although the anaesthetic had not long worn off, Reg recognised his wife and began raising his hand to remove his mask.

'Just speak through the mask,' she reassured him. She was nevertheless delighted to see that he could move his arm. Surely this was an excellent sign.

Over the following days Reg continued to make progress. Sometimes there were severe headaches and sometimes he was more peaceful. Sometimes he spoke sensibly and at other times he rambled. Much of the time Griselda took his hand in hers and simply sat by his side without speaking. She willed him to get better.

Ten days after the operation he was definitely improving. The bruising around his eyes had disappeared and for short periods he was able to dispense with the mask he had been wearing to shield his eyes from the light. He was even asking for a mirror to check out his appearance. With his fingers he could feel the raised seam running from his right ear to the top of his head and he wanted to see for himself how he looked with his right side bald and his left side hirsute—thin though it admittedly was. Griselda forbade it. Gazing at one's reflection was certainly taboo in the facio-maxilliary wards she had worked in during the war, where even the windows were frosted. She saw no reason to make an exception now.

'Your scar is one of the neatest I have seen,' she told him, 'but your hair stylist should be shot.'

'Well this hair cut will certainly cost plenty!' he replied.

Even his sense of humour was returning, she noted with pleasure.

Now that the immediate danger was over the long period of convalescence could begin. Griselda had been warned to expect Reg to be in hospital for up to three months after the operation, and at least four months off work. Her husband would be a difficult patient, she knew. Long, lazy days were just not his style.

The Caribbean cruise was off. Griselda had mailed the confirmation the very day Reg had suffered the stroke. 'Funny, we seem to leave everything just a bit too late,' she wrote to her sister. Likewise the filming Reg was supposed to do with the Australian Broadcasting Corporation. He had been engaged to host a David Attenborough-style television series on the evolution of Australia, with the working title of 'The Road to Now.' They were due to start shooting in November.

'Reg of course is very disappointed,' Griselda wrote to the producer, 'however, for a month or so he will have to slow down. I have tried to impress this upon him for many months—or even a year—but of course his answer is that I should take my own advice!'

On the 14 November, two weeks after the operation, Reg was allowed to go home. Although he was still suffering considerable headaches, the surgeon was pleased with progress and willing to release him into Griselda's care. As usual, she kept her sister informed of progress.

'We seem to be over the worst, though the convalescence of one such as Reg is going to be trying. He is worried that his concentration waxes thin—mine started doing that years ago!

'Fortunately Arkaroola does not need me so I shall be able to make up for all the time we have spent apart in the last ten years, in the next few months. This might sound crazy but I have seen more of Reg in the last fortnight than I have in the previous four months.'

Griselda shepherded her man back to their flat by the sea where he could inhale the salty air and soak up the spring warmth. She ministered to his needs and protected him from the increasingly staccato-like penetrations of company, government and committee affairs with a fierce Scottish impersonation of the fearful Anatolyan wolf hound. She screened visitors and vetted phone calls. Nothing was *so* important. No one was going to jeopardise Reg Sprigg's recovery on nurse Griselda's watch.

It was an approach that worked until Reg, himself, was ready for more interaction with the outside world.

'Dear Graham,' he wrote to the Executive Director of the Australian Petroleum Exploration Association in early December. 'The brain surgeons have drilled into my lid and pronounced a dry hole. I am now on the convalescent list until at least January 1979, by which time I hope to be back on a reduced schedule. I am coming good and giving Griselda hell. I should be at the next January APEA meeting full of the normal bullshit. According to the specialists I have high hopes of complete recovery if I obey the rules. Even the hole in the head is slowly being covered by my massive hair growth,' he joked, with tongue in cheek.

Despite his light-hearted repartee, however, like Griselda, Reg knew that his medical condition needed a considered response. He heeded his doctors' advice and unburdened himself of some of the extracurricular activities that had fueled his pressured life. He asked to be released from the National Energy Research Development and Demonstration Council and resigned from the council of APEA. He had been the founding chairman of the organisation back in 1959, and served continuously in one role or another for the past 20 years. During his seven years as President he had grown the association into the recognised industry representative that lobbied government policy-makers, hosted an annual conference and published its own scientific journal. It was time to take a back seat and let some of the young bloods have a turn.

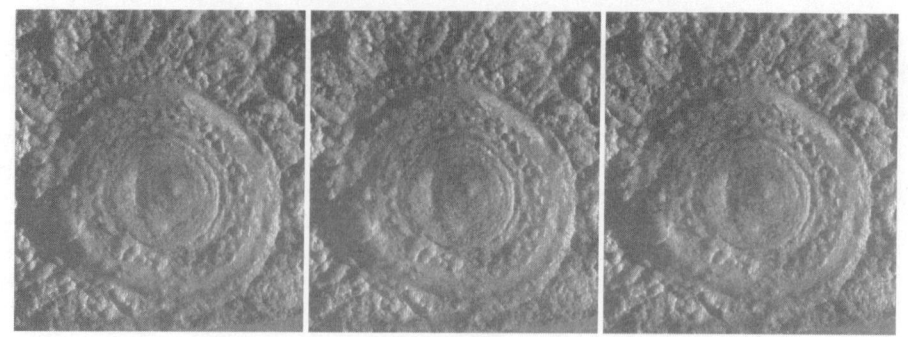

Eoporpita

CHAPTER 12

BATTERED BEACH

AFTER THE JOLT of Reg's aneurism, the 1980s started on a good note. Perhaps, like Reg's friends and immediate family, the professional geological community realised how close they had come to losing one of the most charismatic and intuitive geologists Australia had ever seen. Nomination letters did the rounds, and over the next few years, public accolades washed up at his feet. Being now in his early sixties, it was the time of life when career achievements were being counted. Far better to have Reginald Claude Sprigg making an acceptance speech, flamboyant and full of life, than the wistful regret associated with a posthumous award.

Besides, Reg was renown for being an entertaining speaker. His public lectures over the years routinely drew packed audiences of standing room only. People spoke of his habitual prank of inserting shots of an arty, female nude in amongst his geological picture slides, to give the audience a boost and startle the drowsy. His smutty, schoolboy sense of humour always enervated the talk, and oh, how he drew diagrams of the world—people always remarked on that. The way he addressed the blackboard and whipped out a perfect chalk circle or a map of Australia with a theatrical flourish, always elicited an abrupt round of applause. He was an academic showman who made science fun. And the way he could tie it all together!—the oceans; the land; the heavens; sociology and geography; geology and biology; history and mythology. He wound it all into a narrative that brought the world alive and filled his listeners with the wonder of the universe.

In August 1980 he was awarded an honorary Doctor of Science from the Australian National University.

'Mr Chancellor, it is obvious that Reginald Sprigg's scholarship has had ramifications in a large number of areas,' the Dean of the Faculty of Science announced at the conferral ceremony in Canberra. 'It is not uncommon to find a scientist become an administrator, nor an administrator become an entrepreneur, nor an entrepreneur become a conservationist, nor a conservationist become an adviser to government. But Reginald Sprigg has filled all these roles, and filled them with such flair and verve that he has come to command the attention and respect of geoscientists throughout the world. He has not sought honours, and the Australian National University is pleased that he is accepting the award of an honorary degree at this ceremony today. It is therefore, with great pleasure that I present to you Reginald Claude Sprigg, that you may confer on him the degree of Doctor of Science *honoris causa* on the ground of his distinguished creative achievement as a scholar.'

Even for a man who did not seek honours, it was a satisfying moment. Now he could rightfully call himself Dr Sprigg and put the letters DSc after his name. It helped to alleviate some of the bitterness that still rankled over his own aborted DSc at the University of Adelaide all those years ago.

But not completely. He never gave up believing that others had conspired against him to besmirch his efforts to legitimately obtain the academic qualification back in 1952. To see current researchers reinventing his early

ROCK STAR

Reg receives the degree of Doctor of Science *honoris causa* from the Australian National University, 7 August 1980. Opera singer Dame Joan Sutherland received an honorary degree of Doctor of Laws for 'distinguished creative contributions in the service of society,' at the same ceremony.

theories added salt to the wound. Perhaps he would try to untangle the mystery of what really happened, one day.

Now, however, was the time for enjoying this unexpected honour. And—for a music lover such as himself—to be sitting on the stage of Australia's most prestigious university next to the opera singer Dame Joan Sutherland. What a thrill! She was receiving an honorary degree of Doctor of Laws and Joan and Reg beamed at the audience and at one another with delight.

For Reg, having an award from this particular institution also formed a spiritual link to his dear friend Sir Mark Oliphant, which he treasured.

It was 33 years ago that they had first met, in the parched summer heartland of Mount Painter. The wilting Cambridge university professor and the eager young geologist who brought him a drink. In the years since, they had pursued their respective scientific careers; their joint interest in uranium providing a point of reference for their paths to cross from time to time.

In 1950, Mark Oliphant had interrupted his flourishing career in the UK to return to his homeland and spearhead the founding of Australia's national university. He dedicated his remaining academic years to the establishment of the Research School of Physical Sciences at that institution.

Oliphant was 18 years older than Reg, but their lives had followed similar paths. Like Reg, Oliphant had studied and been a cadet at Adelaide University, and for a short while, they had even attended the same primary school at Goodwood, albeit in different years. There were intrinsic similarities, too, between the two men. They were both instinctively friendly and outgoing; enthusiastic and energetic; and interestingly, neither was cowed by authority. Reg was never reticent about sparring with Mawson on scientific differences of opinion and likewise Oliphant had engaged his physics professor, Kerr Grant, in forceful debates. Students daring enough to challenge the almighty professors were a rare and remarkable commodity in the strait-laced era of the 1920s and 1930s.

Above all, Sprigg and Oliphant both shared an intellectual curiosity that shaped their personality and imbued their entire being. It was the over-arching definition of who they were. The kernel of their humanity. For both these men, the free activity of the mind was the meat and drink of life itself.

'The important thing about men like Oliphant is not that they perform brilliantly in conventional academic exercises when young, but that they develop extraordinary lateral thinking and innovative powers,' recalls one of Oliphant's fellow university students in Stewart Cockburn's and David Ellyard's biography *Oliphant*. 'These enable them to look at known facts, arrange them in some new context, and help them to perceive new truths hitherto unrecognized by others.'

Reg Sprigg, too, was one of those men.

In November 1971 Oliphant had become the new Governor of South Australia. His five years in Adelaide's Government House had given him the opportunity to renew acquaintance with Reg Sprigg. He had always liked and respected Reg and after a vice-regal visit to Arkaroola in 1972, his admiration had grown stronger. Oliphant's childhood memories of long family walks in virgin Adelaide Hills bushland had bred in him a strong environmental bent and he applauded anyone who toiled to protect such an irreplaceable resource. From then on, the lines of friendship between the two drew tighter each year, strengthened by their scientific companionship and their joint love of Arkaroola.

When Oliphant learnt, in June 1983, that Reg had been made an Officer of the Order of Australia in the Queen's Birthday Honours, it reminded him of receiving his own Imperial Honour back in 1959.

'Dear Reg, our warm congratulations on the award of the AO,' he wrote to his friend. 'It should be a knighthood, for then—as Slim said when he dubbed me—it is shared by the wife, and Griselda must surely take some of the credit. Mark'

No doubt, Griselda, would have been ecstatic with the idea of Sir Reginald and Lady Griselda Sprigg. As much as Reg downplayed the accolades and modestly reminded people that he was part of a team and owed much of his career success to support from the wider geological community, his wife revelled in the reflected glory and proudly displayed Reg's certificates and awards in the restaurant at Arkaroola—much to his embarrassment.

The 1980s also saw Reg made a distinguished member of the Petroleum Exploration Society of Australia for, 'exceptional and meritorious service rendered to the Petroleum industry of Australia;' a fellow of the Royal Meteorological Society; and an Honorary Fellow of the Royal Society of South Australia, 'in recognition of distinguished services to the Geological Sciences in Australia.' It was the first society he had joined, as a 17-year-old schoolboy back in 1936. He was also awarded the inaugural Lewis G Weeks Gold Medal for outstanding achievement in the petroleum exploration industry, and later would come the granting of the Freedom of the City of London and another honorary Doctor of Science, this time from Flinders University of South Australia. He was gathering quite an alphabet of letters to put beside his name, if he so wished.

ON THE WORK front things were also progressing well. The Colson well—that Reg had been planning to visit at the time of his cerebral haemorrhage—had proved dry, like so many others. Four years, $1.6 million, and no oil. But Beach had at last had a win, with its first commercial hydrocarbon discovery.

In November 1979, in south-western Victoria, Beach's North Paaratte-1 well had struck gas. Seven million cubic feet of it bursting up through the well head each day. It seemed that—finally—Reg's faith in the Otway Basin had been rewarded.

The North Paaratte well was on land that had been repeatedly tested over the years. The Frome Broken Hill group in a joint venture with Shell, first surveyed the area in 1958 and struck a promising gas flow at Port Campbell the following year. At the end of a week-long production test, however, the gas petered out. Hydrocarbon shows in almost half of the 45 wells drilled in the Otway Basin over the following years teased the major oil companies enough to keep them interested in the Victorian half of the basin, while Beach persisted in the South Australian half. All were hopeful of a recurrence of the nearby Gippsland Basin structures which fed the enormous Bass Strait oil and gas fields to the east. But finally, at the end of 1976, Shell declared enough was enough.

'The Otway Basin is dry,' Shell's general manager informed the Melbourne *Age* newspaper. The prospect of the Victorian territory holding commercial reserves of oil or gas—either on- or off-shore—was virtually nil, he declared, and after spending more than $15 million over the past ten years, he was calling off the search.

Reg remained steadfast. When Shell surrendered their Victorian acreage, Beach took it up and immediately began re-analysing Shell's seismic survey data. Beach geologists suspected that their predecessors had been looking too deep and that if hydrocarbons were trapped in the area they would be found in the shallower sections, such as the Upper Cretaceous Waarre Sandstone. With its first well Beach found success at 440 metres.

Mindful of the failed Port Campbell well nearly 20 years earlier, Beach was cautious in its announcement, but investors needed little convincing and forced Beach shares up more than 25% on the day the news became public. This time their optimism was rewarded.

The existence of a commercial gas field was confirmed when the North Paaratte-2 well and nearby wells at Wallaby Creek and Grumby were drilled early in 1981. Beach estimated that it had 15.37 billion cubic feet of natural gas in the ground, enough to supply the nearby city of Warrnambool for the next 20 years. It was a relatively small gas field but a major milestone for the company. Agreements were signed, a production plant designed and gas deliveries commenced in April 1986.

At about the same time, in June 1984, a drill rig pierced an oil pool at Bodalla South in Beach's Cooper/Eromanga Basin lease area in western Queensland. Within months Bodalla South crude was flowing to the Eromanga refinery and Beach received its first ever petroleum income—a cheque for $27 783.01 representing Beach's 20% share of the first 6615.04 barrels. After 23 years of effort, it seemed that the company's fortunes had finally turned.

Two years later, as Beach celebrated its 25th birthday at the end of 1986, both gas and oil income were flowing into the company's coffers and Beach posted its first profit.

'At the commencement of the new financial year, Beach is an oil and gas producer, is free of debt, has net current assets of approx $14.5 million, has the support of two major shareholders holding near 50% of issued capital, and holds some exploration acreage,' the Chairman wrote in his annual report to shareholders. 'We are confident that not only will the company develop as a petroleum producer but will be in an excellent position to take advantage of good opportunities that the current downturn may provide.'

Approximately 4500 barrels of oil per day were being produced at the Bodalla Block, which now encompassed two neighbouring oil fields. Despite a downturn in crude oil prices from net proceeds of $33 per barrel at the beginning of the year to $6 per barrel seven months later, Beach posted a $6 million profit, and the following year—1987—was able to pay a maiden dividend to shareholders of 1.5 cents per share. Twenty-six years was a long time to wait for a meagre return on investment but in this respect, Beach shareholders were not alone. It had taken Santos just as long to return its first dividend to shareholders, who had in 1978 pocketed 2 cents per share, 24 years after the company was incorporated.

At the close of the 1986/87 financial year Beach reported that sales of Bodalla crude were up 10% from the previous year; gas sales from the North Paaratte field were 14 times higher than the previous year, as the first full year of production was realised; and despite the slump in oil prices, the company had posted a second consecutive annual operating profit of $6 million.

Beach was recognised as a respected junior oil and gas explorer with long-term vision and a commitment to on-going grass-roots exploration. It was a success. It had a good reputation and business appeal. And it wasn't long then, before the independent Australian company looked like a tasty morsel for a corporate takeover.

Business analysts predicted that a takeover offer would come from one of Beach's principal shareholders: North Broken Hill Holdings Ltd, holding 27% of Beach; or Peko-Wallsend Ltd, holding 23% of shares. When it did come, however, the offer was from an unexpected quarter.

In March 1987, Claremont Petroleum (through its wholly owned subsidiary Midland Exploration), made a play for Beach. Like Beach, Claremont was a junior oil and gas company. It, itself, was considered to be vulnerable to a takeover bid and may have decided to 'eat rather than be eaten'. Combining the two companies would create a corporate group that rated ninth amongst Australian oil and gas producers in terms of current production.

The Beach directors urged their shareholders to hold firm against the invaders. They had all rejected Claremont's revised offer of 82.5 cents per share, up from its original offer of 70 cents per share. But when North Broken Hill sold its 26.8% stake, the outcome for Beach was set. By mid-June, Claremont had acquired 65% of 'Reg's' company. He was dismayed.

'It's sad that once a company is successful, has a good income and money in the bank—then the predators move in to snatch the prize,' he wrote to his friend Oliphant.

Changes were immediate. Four Claremont nominees joined the Beach Board and three former Beach Board members stepped down. One of them was the company's founder, Board member for 23 years and deputy Chair for the last four, Reginald Claude Sprigg.

For Reg, it was an appropriate time to go. At Griselda's urging he had actually attempted to resign from Board duties several years ago, in 1984, but had been persuaded to stay by Chairman Alan Lobban.

'I am most reluctant to accept the tender of your resignation,' he had written to Reg. 'I am the first to acknowledge the long and most valuable service you have given to the company but I do not accept the proposition that now is the time for you to "retire quietly". I believe that your advice and assistance during the coming months will be more valuable than ever and I for one would miss this very much. I am sure this feeling is shared by my colleagues on the Board.'

Although his position of consultant technical director had been reduced to one week per month in Melbourne—timed to coincide with monthly Board meetings—rather than the two weeks per month he had originally agreed to, the strain of the constant travelling was growing more and more tiresome. Especially now that he and Griselda were living permanently at Arkaroola, and Beach had moved from offices in central Melbourne to suburban Camberwell. It all meant an extra two days driving at the Adelaide end and extra rigmarole at the Melbourne end.

Since having quadruple heart bypass surgery in November 1985, Griselda refused to let him make the drive down from Arkaroola on his own, so it was an imposition on her too, unless he could fly or take the bus. Sometimes he wondered if his contribution to company affairs was really worth all the expense to Beach and the physical exertion to himself. He was 65 after all, and wasn't this a time in life when one was supposed to start taking things a bit more sedately. Would that time ever come, he asked himself. And would he actually enjoy it if it did?

At the time, Griselda had grudgingly allowed him to continue duties with Beach, even though she knew that Board weeks in Melbourne were a strain for her husband. Back in 1978 he had been warned by his specialist to change his lifestyle but so far she had failed to see any difference. She felt she had spent more than 30 years of marriage living as a single person, and she was sick of it. One year, she remembered, she had counted up that he was home for only 22 days. And none of them consecutive! Now—they both agreed—it was time for Reg to stop.

At first he just withdrew from Board duties. Six months after the takeover he retired completely.

He was followed soon after by the departure of John Hinkins, Beach's Executive Director for the past 11 years, and the transfer of all staff to Sydney, where Beach management would be integrated with that of Claremont. Only eight Beach staff made the move. It was a changing of the guard. The old Beach was gone and a very different creature was emerging in its place.

As Reg watched from the sidelines he became increasingly irate. Despite Claremont's pre-takeover assurances that Beach would continue as an autonomous oil and gas company, Reg was suspicious. He didn't trust the new owners.

'They are *not* interested in exploration—only company manipulation!' he wrote indignantly to his friend Oliphant. He ranted about the, 'devastation caused by lawyers and accountants taking over. They've all but ruined a company at last getting its head above water.' This, only nine months after the takeover. Perhaps he had a premonition that there was worse—*far worse*—to come.

Behind the scenes there was a tangle of creative sharetrading taking place that culminated in a predatory outfit called the Independent Resources Limited (IRL) Group gaining control of Claremont and thus Beach, without even making a formal takeover bid. Michael Fuller, an Adelaide lawyer and IRL director, was appointed to the Boards of Beach and Claremont in mid-1988, followed a few months later by fellow lawyer and IRL director, Joseph Cummings, and yet another IRL director, Paul Main, in mid-1989. And so the free-fall began.

The IRL directors with their hands on Beach's reins, launched an immediate 'reassessment of the Group's exploration programme in Australia and New Zealand.'

After pocketing earnings from gas sales of $1.2 million in the 1988/89 financial year, Beach's North Paaratte gas plant was sold in 1989. Six exploration permit areas were relinquished because they were considered, 'high risk and therefore unattractive to Beach, compared to other opportunities available elsewhere.' The permits included acreage in the Eromanga Basin, a few kilometres north of Beach's producing Bodalla South oil field; two areas in the Surat Basin and one area in the Clarence/Moreton Basin in south-east Queensland; Beach's entire acreage in the Pedirka Basin in the Northern Territory Simpson Desert, covering the Colson, Poeppel Corner and other drill sites; and part of the Otway Basin in Victoria.

The sell-off and curtailment of its exploration programme boosted Beach's available cash and deposits from $23 million in mid-1987 when Claremont took over, to $39.7 million by mid-1989. One year later, by the end of June 1990, these assets had been reduced to a measly $4 million.

In lieu of the Australian permits deemed 'too risky' to explore, the new IRL directors had taken Beach's bank-books overseas and for $38 million, purchased the entire issued share capital of a Liberian-registered company, Mazeley Ltd, and its interests in gas and oil fields in Oklahoma, USA. Apart from the complicit IRL directors and advisors, no one at Beach could have possibly known then, that the vendors of these Burbank fields were also IRL-related companies. When Beach didn't have enough money to pay the entire amount three more IRL companies were happy to advance Beach a loan—the so-called Syndicated Cash Advance Facility or SCAFA. Beach was now a fly in the IRL web and it had all happened in the blink of an eye.

Jean Shaw, a loyal Beach shareholder since 1964, remembers the meeting in April 1990, when she first sensed that her company was in some kind of trouble.

'The first I became aware there were problems was when trading was suspended—the shares weren't appearing in the paper,' she recalls. 'I rang the ASX to find out what was going on and I was told that until they handed in an Annual Report they wouldn't be on the market.'

In fact, Beach had been suspended from trading since November 1989 for failing to complete its annual accounts and annual report, and was being threatened with delisting from the Exchange unless a series of written questions concerning the US oilfields deal was answered. Jean visited the Stock Exchange in Adelaide the day before the postponed Annual General Meeting (AGM) to gather the facts.

'They armed me with some of the things that were going on so I could ask questions in the meeting. But they didn't answer anything truthfully,' she says. 'I could tell they were lying through their teeth. So I thought, I'll get hold of the minutes of the meeting and then have some ammunition. So I applied for a copy, and discovered there weren't any…They'd been shredded.'

Reg was another shareholder demanding answers, admitting he was 'dreadfully concerned that what was a very respectable company is now in a dubious state' and financially appeared to be 'skating on thin ice.' At Claremont's five-hour meeting the day before Beach's AGM, Bruce Beeren, a general manager at the Australian Gas Light Company, which had a 20% stake in the company, went even further, publicly declaring that Claremont had been 'raped' during the two years in which Michael Fuller had been in charge. An accusation which Fuller roundly rejected.

However, when the 1990 annual report came out at the end of the year, revealing the extent of the companies' losses, Beach and Claremont shareholders were horrified. Beach had posted a net consolidated loss after income tax of $15 million. The healthy financial assets of one or two years before had been squandered.

Claremont minority shareholders (including Westpac and the Australian Gas Light Company) immediately called for an Extraordinary General Meeting (EGM) to remove their Board. At the November AGM, which was scheduled one week before the EGM was to take place, they denied Cummings's re-election and appointed two new independent directors. They also signalled that Fuller's upcoming bid for re-election to the Board would not succeed. The day before the EGM, he resigned. When voting was over the next day, Claremont—followed by Beach two months later—had an entirely new Board, totally independent of the IRL group.

The first job for the new Board members was a complete assessment of each company's plundered assets, including a detailed examination of the American oilfields now owned by Beach. Within a fortnight of their appointment, two directors were in the United States on a fact-finding mission.

The Burbank Oilfield, north-west of Tulsa in northern Oklahoma, had begun production in the 1920s when it was judged to contain 671 million barrels of oil. More than 2500 wells had since been drilled across the field. But it was obvious that the oil couldn't keep flowing forever. In the 1950s, primary production had been replaced by a secondary waterflooding recovery technique necessary when there is insufficient pressure in the underground reservoir for the oil to rise to the surface by pumping alone. By the 1990s, when Beach bought in, tertiary production techniques were being tested in a bid to withdraw the last remaining barrels. Beach had bought into an oilfield on its last legs.

Beyond the fact that their new oilfield was a lemon, the most immediate concern for the directors in January 1991, was how to service Beach's outstanding $5 million loan from IRL-controlled Jingellic Minerals, for the purchase of the Burbank field. As the Board members investigated every detail of the deal they came to the conclusion that the entire transaction had been designed to defraud both Beach and Claremont, and thus they refused to make any further repayments. Jingellic took Beach to court to have the company liquidated. Beach mounted a spirited defence and in turn took Fuller and Cummings and the whole damn lot of them to court and vowed to fight the corporate raiders to the bitter end.

At Arkaroola, Reg followed the proceedings with rising fury. He had always despised the corporate cowboys—the Grenfell Street miners—who used companies to line their pockets with never any thought of serious exploration. From the 1950s—with Santos—til now, nothing had changed. But in his view this was the worst of it, by far!

He had known something was wrong right from the start, he berated himself; right from when Beach was first snapped up by Claremont. Then, when those IRL boys came on board…It was an ominous sign, he believed, that they each had held only a paltry 1000 Beach shares. Why, when he'd been a director he'd had at least 70 000 shares in Beach, showing a concrete faith in the company's prospects. And Griselda was always listed amongst the 20 largest shareholders, with her 217 160 shares—0.2% of the company.

While there was nothing he could have done to save Beach from the maelstrom, he nevertheless rued the whole stinking episode.

But as was Reg's way, he didn't languish in despondency. He and Griselda were now living permanently at Arkaroola, and Arkaroola was his primary concern.

Building of their home there had proceeded in fits and starts. It had been three years in the making and more than 12 years since he had first proposed the idea to Griselda in 1969, not long after they'd bought the property. Since it was really somewhat of a luxury to build a new dwelling exclusively for their own use, construction could only proceed when all other bills had been paid and there was a little money left over to buy windows, door frames, roofing materials, bricks or what have you. Cajoling the builder to work when money and materials were available and then halt construction when funds ran out was tricky, but they'd never envisaged that it would go that way. As usual, problems that diverted money to more pressing needs, cropped up when least expected.

But finally, by the end of 1981 their house was finished. It was a long narrow building with a low profile so that it tucked discreetly into the base of the hill behind it. A scrubby, rocky slope angled upwards just metres from the rear of the house. Sometimes euros came down to forage around the house and you could see them take off up the hillside from the kitchen window, bouncing swiftly over the rocks. Reg detested small pokey rooms so the kitchen, dining and spacious living room melded into one open area fronted by floor-to-ceiling windows that afforded a long view down over the village and out to the hills beyond. If she could steal some time away from the motel or Reg's typing, Griselda could lean back on her couch and puff away on her cigarettes with the world of Arkaroola at her feet; Reg scribbling away behind her and his classical music pervading the house. Just a few minutes there could revitalise her. This, surely, was what they had worked so hard for, over so many years.

Initially, back in 1978, when the house was just a great concrete slab, and had stayed that way for many months, Griselda expressed some doubts about the wisdom of building a large home for themselves. She had so little time for housework that the 'doll's house' they inhabited when at Arkaroola seemed more than enough for her to look after. But Reg insisted on somewhere a little more remote. There, they were right in the midst of the village and there were constant intrusions as staff and sometimes even guests, popped in 'just for a moment.' Interruptions caused by maintenance emergencies or certain motel problems were unavoidable, of course, but the endless stream of passers-by eroded his writing time and became a source of increasing frustration for Reg. He was never able to turn people away and each guest was made to feel that their visit had been the highlight of Reg's day, but privately he longed for some hours of his own—time that he could devote to his writing.

Their new house on the hill, which Griselda named Ennismore, after the Scottish hotel in which they had first met, gave him some distance. The five-minute walk from the motel made him not quite so accessible, and there, with his music, he could write away the hours. At the long dining room table, with papers amassed over table and floor, he would churn out hundreds of hand-written pages.

In her study at the other end of the house, Griselda would polish her glasses, decipher the scrawl and type up the pages in her two-fingered way. For Reg had decided to write a book, an autobiography of sorts, with lots of geology thrown in, naturally enough. He felt that he had been fortunate to have been involved with lots of interesting projects throughout his career—met lots of fascinating people and been privy to some important discussions—and now it was his task, nigh his duty, to record it all for posterity.

He had already written two books on the natural history of Arkaroola. This would be different. A collection of humourous anecdotes and episodes spanning his career and the geology, *always the geology*, that had led him onward. Being honoured with a string of awards, especially the Order of Australia, had stimulated his gratitude for those along the way who had motivated and encouraged him.

'I would that "Dougie M" were alive,' Reg wrote to Oliphant. 'He was an inspiration to me—as was Madigan, Howchin etc. I was a lucky young enthusiast.

'Now all I want to do is to get around to writing, *Geology is Fun*,' as he had decided to call his first volume of memoirs. 'We geologists are a funny breed and it's good to laugh at oneself from time to time.'

Oliphant encouraged him to get it all down on paper or tape. 'You have lived a remarkable life, through remarkable years, and it is very important that all should be on record, warts and all,' he agreed with Reg.

Oliphant wasn't 'speaking through his hat,' either. He himself had gone through a similar process, being the subject of a biography published in 1981. From experience he knew that it could be a harrowing process—offering your entire life up for public scrutiny, but he always had been one for transparency and scrupulous honesty, and he believed in a complete historical record. Reg, he knew, was of the same ilk.

'You are a thoughtful person, aware of the fragility of life and the importance of *correct* history,' he wrote to Reg.

The correspondence between the two, which had begun in earnest when Oliphant had retired as Governor and Reg had finished with Turkey, was a weekly occupation for Reg. Besides his other writing he always had masses of letters to attend to, speeches to compose, scientific publications to compile. Perhaps if he was not such a profligate letter-writer himself, the return correspondence would not be so voluminous but it was a compulsion for him and he knew no other way. He wrote letters—quite often two or three or four—every day of his life, Christmas included. While he admitted that sometimes it did become tedious, one thing he never tired of was corresponding with Oliphant.

'You apologise for your poor efforts as a correspondent. Quite the reverse. I find you one of the few who inspire me these days,' Reg wrote to his friend in the midst of the Beach/Claremont fiasco. 'I need more of such people's stimulation. I wish you weren't so far off.'

Their letters back and forth covered the gamut of subject matter from history and technology to alternative energy, the incredible advances in molecular biology, and conservation of the planet. A favoured subject for Reg in the late 1980s was the philosophical question of Gaia—the theory that the earth had some kind of innate balance mechanism, a type of feedback system that somehow maintained planetary conditions in a biologically favourable zone. It was thought-provoking stuff; an interesting idea that was almost a scientific alternative to the religious creationist mantras. To Reg it did conjour up the image of an omnipotent overseer; a 'God' for the scientists, if you like.

'I'm intrigued at the extent to which the idea of GAIA is permeating so many branches of scientific thinking. I wish I could evaluate it far better,' he wrote to Oliphant. 'The more I read (and even think) the more I can appreciate that "something" seems almost to be guiding evolution of our earth—but with many fits and starts. Why should plankton "bother" to make dimethyl sulphide which evaporates and becomes oxidised in the atmosphere to form a type of sulphate "aerosol" which then becomes of benefit to creating and/or maintaining favourable conditions for life on earth? Certainly earth's maintenance of an atmosphere at zero to 30 °C down the last several billion years is remarkable. Needs all tricks to achieve it.

'The realization that major proportions of earth-bound life are literally wiped out almost regularly (every 26 million years or so), is now exercising many minds also. Maybe a nuclear wipe-out of higher life forms on planet earth as we know it is not such a catastrophic concept after all. Is it possible to conceive of a "higher energy," silicon-based, life form on another planet in another universe—rather than our carbon-based existence?

'All quite incredible,' he concluded. 'Enough to make a philosopher out of the most mundane.'

Reg loved to occupy his mind pondering such things and he was grateful for someone to debate with, although Oliphant was more pessimistic than he. The physicist's intimate involvement with the development of the atom bomb, and then his absolute abhorrence of the human destruction it had wrought had adversely shaped his opinion of human nature. He felt more and more despondent at mankind's inability to curb its technological appetite. No good would come of it, he was sure.

While Reg shared Oliphant's concern for the ultimate future of the human race, he was more optimistic that people's inventiveness would prevail and that benefits too, could be achieved through technology. As long as we didn't destroy the earth's riches first, he warned.

'I see by an article in the latest journal that the CO_2 Greenhouse effect is not appearing so rapid—not until 2030 do they expect serious melting of the ice caps!!' he wrote, incredulous, to Oliphant. 'Surely now is the time to take more drastic action *before* it is too late. We seem determined to mortgage the future, making it easier for us right now. Enjoy now, pay later.'

He revealed his habit of standing on street corners in Melbourne while awaiting the tram, and counting the number of occupants in each passing car. At around nine o'clock in the morning approximately one in every 13 cars carried a passenger, he told Oliphant. By late morning he estimated that only one in every 20 or 30 cars contained more than one occupant.

'Such an incredible waste of fuel,' he lamented. 'God, we Westerners are wasteful of resources. Will we last to the end of the century? How can we do without atomic energy as our fossil fuels run out; is nuclear fusion a real potential before it is too late? Is an annihilating nuclear winter a real possibility if the super-powers ever get to a nuclear-age conflict? I wonder.'

While there were some topics that the two scientists didn't completely agree on, there was one topic on which they were in lockstep—their shared love for Arkaroola. When Reg felt too downcast about the problems of the world he could step out on to his slate verandah and inhale the clean dry air of the place. He liked to measure the day at his home weather station and absorb the rusty ambience of the rocky outcrops all around. When he could steal a few moments he would roam the property on foot or do some low, slow flying with Doug in his Auster.

'God this wild country is magnificent,' he wrote to friends. 'Drifting among the peaks and deep gorges is but a dream. I never thought I would find it so enchanting and exciting. Cool days, deep shadows: the solitude and grandeur of it all, all but defies my humble description. I love the place more every minute.'

It made world politics and international conflicts shrink into oblivion, and it made Reg Sprigg feel like one incredibly lucky human being.

SINCE OLIPHANT HAD visited Arkaroola as South Australian governor in 1972 and seen the private sanctuary that Reg had created, he had wasted no opportunity to revisit with the Spriggs and drink in the splendour of the northern ranges. Hiking among the wattles and eucalypts transported him back to a boy of eight years old, when he would walk 10 kilometres each day to and from his home near Mylor to the little one-teacher school. He also relived camping holidays with his father and younger brothers where they walked from the hills all the way to the south coast, sleeping in the open and living off the land. Simple joys that had given a lifetime of pleasure.

Now, exploring the Arkaroola countryside awakened in him the same feelings of deep contentment. He found its hills, its valleys, its arid landscapes, and its wealth of changing shapes and colours, uplifting. For him, it was 'solace for the soul of a dreamer.'

In 1986—the year of South Australia's 150th jubilee—Reg decided to honour Oliphant and his connection with Arkaroola, in two ways. The first, was the official naming of Mount Oliphant, a stout, rounded peak about 5 kilometres north of the village. The second, was inviting Sir Mark to open Arkaroola's new observatory. Reg, and especially Douglas, had had a lifelong interest in astronomy. Arkaroola, with its clear, dark skies was the perfect location for stargazing. On a peak on the western edge of the village, near Reg and Griselda's house, a little igloo-shaped structure pugged together with local stone and cement took shape. A smooth, rotating tin dome with a wedge-shaped roof opening went on top and inside went a 150 millimetre telescope (later upgraded to a 360 millimetre Celestron C14).

It was ready by the 18 March 1986, just in time for Halley's Comet's one-in-76-year close approach to earth. Appropriately enough, 84-year-old Oliphant was the only one of the guests attending the occasion who was

View from Griselda Hill down to Arkaroola village.

seeing the comet for the second time. He had been a boy of eight years old when it had last passed close to earth in April 1910.

Reg had decided to name the observatory in honour of George Fredrick Dodwell, South Australia's government astronomer from 1909 to 1952 and pioneering geophysicist. He was a man largely lost to modern consciousness but Reg had known him briefly in the 1930s and admired his early magnetic surveys and his attempts to stimulate public interest in the heavens. His fondest memory was of Dodwell parking his car behind the Geology Department and racing in to Mawson, calling, 'The discovery of crab-like fossils on Mount Everest proves the Bible correct! The Great Flood did cover the Earth.'

Mawson distracted Dodwell's attention long enough to slip out to the car, where he—typically injudiciously—told Dodwell's wife, 'Your husband's gone off his head. He is mixing geology with theology and getting it all wrong.'

IN ADDITION TO the observatory, Arkaroola's facilities had now grown to include a third 10-room motel lodge, plus a swimming pool, playground, pioneer/folk museum and sundial park. The dining room and motel reception complex had been refurbished inside and out, and bitumen had been laid around the village at a cost of $26 000, to reduce the dust.

Reg supervised the construction of all these additions and also pottered with his own private pursuits. He was never idle. His interests were as vigorous as they were varied. He created metre-long mosaic pictures from one centimetre-square, multi-coloured ceramic tiles and propped them along the entrance road to the village. He constructed two large concrete murals—one depicting sea floor spreading and the other an Ediacaran beach scene—and cemented them into the front corners of the main building. He designed and established a Mesozoic garden to recreate the landscape of the dinosaurs at the entrance to the village's visitor information centre, and carted and arranged 70 tons of colourful stones to create a giant depiction of the Ediacaran worm *Spriggina*—an ancient marine animal with the superficial appearance of a centipede, which used its sinuous body and multiple legs to skitter along the ocean floor beneath the warm, calm seas. The stone model was 70 metres long—1500 times larger than the four- to five-centimetre-long fossils of the real *Spriggina*—and best viewed from above a nearby lookout.

It was in June 1958 that Adelaide University's Dr Martin Glaessner had named one of the quirky Ediacaran animals after Reg. The lecturer had just returned from his first expedition to the area and was enthusiastic about what he'd found. It was almost 10 years since Reg's discovery. After his first two trips to Ediacara in 1946—when he had made the first find and then confirmed it with several other specimens—almost nothing had occurred in the intervening years. Reg had made a third visit in mid-1947 with his then-wife, Pat, and palaeontologist Curt Teichert, which had culminated in the romantic affair between the two. It did dampen his enthusiasm for all things Ediacara for a while.

By March 1948 he was off on his nine-month tour overseas. Showing photographs of his fossils at tea at the British Museum elicited some polite discussion but an entreaty for a formal presentation to the 18th International Geological Conference then meeting in London was unsuccessful. There seemed a reluctance to credit the find to such ancient rocks. Reg found the general apathy amongst his professional peers deeply discouraging.

When he returned to Australia there was Radium Hill to manage and the new Department of Mines mapping division to establish; a new wife and home to organise; a baby and his own company to raise. Time off for fossil hunting was virtually non-existent.

It wasn't until 1957 that interest in Reg's Ediacaran fossils was revived by a local school teacher. Hans Mincham had taught at the Beltana school, nearby to the site, and knew the area well. He was a natural science buff who had visited the old Ajax mine a few kilometres north of the town looking for *Archaeocyatha* and had also been out to Ediacara, searching for insects. When he read of Reg's discovery in the *Transactions of the Royal Society of South Australia* he determined to make another visit there.

It was December 1956 before he and friend Ben Flounders—a cabinet-maker and keen amateur geologist from Whyalla—made their excursion to the red hills of Ediacara. Not surprisingly a heatwave coincided with their pre-Christmas visit and fossil-hunting was cut short. Rocks lying on the ground were too hot to hold and the fun of the trip evaporated with the sweat on their backs. However the couple of eroded impressions they found were promising enough to stimulate a second trip in the September school holidays of 1957.

This time the weather was perfect and so were the pickings. Walking back and forth in the low-angle light of early morning and evening Ben and Hans gathered a fine crop of fossils. Many were representative of those that Reg had found before but others were completely new, including the segmented marine worm with a curved-arrow-shaped head and a curious little shield-shaped creature with a ridge on its back like an anchor.

When the two amateur naturalists presented their collection to the South Australian Museum, official interest was finally ignited. The museum board approved a special expedition to make a thorough and methodical search of the Ediacara site in March 1958 under the leadership of museum palaeontologist Brian Daily and in the company of Hans and Ben. Six men combed the area for five days and hauled out two truck-loads of material.

Two months later on the 29 May, the Lieutenant-Governor proclaimed the entire site (2500 acres) an exclusive fossil reserve.

By this time, palaeontologist Martin Glaessner, had at last also turned his eyes to the Ediacaran fauna. Martin was known to be one who took a while to make up his mind on an issue and was not averse to performing a scientific back-flip if sufficient knowledge accumulated to challenge his original view. He'd been leery of the 'continental drifters' when theories about tectonic plates moving around and continents sliding under one another first came along, and likewise Reg's first reports of so-called fossilised jellyfish. But now he was convinced the animals were real.

When ANZAAS had their annual meeting once again in Adelaide that year (1958), Glaessner and Daily led a tour to Ediacara. From then on, Martin Glaessner made the study of these ancient soft-bodied animals his life's work. He named the enigmatic shield-shaped critter *Parvancorina minchami*, in honour of Hans Mincham. A jellyfish reminiscent of a repeating tudor rose-type pattern—which Reg regarded as the most elegant of all the Ediacaran fauna—he named *Mawsonites spriggi*. Reg was delighted with the permanent association of his name with that of the man whom he most admired.

The swimming centipede-like creature Glaessner named *Spriggina floundersi*, thus jointly honouring the original discoverer of fossils at Ediacara and Ben Flounders, the one who had first found the marine worm specimen. When, at a Royal Society meeting, Martin introduced *Spriggina* as the lowliest worm that ever lived, Reg countered that its concentration of nervous tissue at the head end surely made it the brainiest creature in the world at the time! They may both be right.

Funisia dorothea

CHAPTER 13

ONE LIFETIME IS NOT ENOUGH

ON THE FIRST of March 1989, Reg turned 70. A couple of hundred friends and family drove to Arkaroola to join him in celebrating. Some of the older guests cancelled at the last minute because they were too frail to travel in the extreme heat of that late summer. The hot, dry air was a meteorological trick, however. Soon after the party—even before the last guest had departed—it started to rain and it didn't stop until Lake Eyre was full. The outback was experiencing a 'once in a lifetime' flood and city-dwellers flocked to the north to see the spectacle. Arkaroola's pilots were kept busy each day flying tourists out over the flooded lake and back. The crusty, blinding white salt pan was gone; in its place a greyish expanse of water with waves and fish and pelicans. What a transformation.

Because of the rain, Arkaroola was having a rare bumper season in two ways: the chronically dry ranges were enjoying a spurt of regeneration, and visitor numbers were up. More than 21 inches (533 millimetres) of rain had fallen on the property between March and July and the country had responded magnificently.

'It looks like the Emerald Isle!' Reg told his friends in England.

Creeks were flowing in a way they hadn't since 1974. At each new downpour the normally tinder-dry Wywhyana Creek touched the 3-metre mark. Boulders weighing many tons bounced and tumbled in the surge. Trees and bushes dripped a coat of bright green varnish. Frogs mingled the unaccustomed sound of their croaking to the exuberant accompaniment of bird calls. The very air sparkled with freshness.

The dirt roads, however, were a consequent mess of furrows, washouts and muddy quagmires. 'Both good and bad news comes to us together,' Reg stated philosophically. But at least there were tourist dollars in the kitty to pay for the damage this time.

When Australia's pilots when on strike later in the year, Arkaroola went to full capacity.

'It seems that all those bodies heading interstate for holidays have got in their cars and headed north instead,' Griselda wrote to her friend. 'We have never been so busy in all the 21 years of Arkaroola's existence,' she exclaimed.

The three motel lodges were full every night and scenic flights were in high demand. Griselda put Reg's typing and company business aside and resumed her former role in the motel, receiving guests and serving in the shop and bar. At 67 years of age, she felt it now—being on her feet all day—but the staff really were flat-out. And Arkaroola always came first.

ALSO IN 1989, Reg published his memoir, *Geology is Fun*, a B5-sized, 349-page, hard-cover book that covered the first 20 years of his geological career. The book represented a considerable investment of both time and money. But he was glad he had done it. Now he wanted to get straight on with writing the second volume—an account of the following forty years that would come out four years later under the title, *A Geologist Strikes Out*. That book would cover the Geosurveys years and beyond; the formation of Santos and Beach Petroleum; and some of the commercial dealings that he wanted to get out in the open.

Santos, in particular, was one subject that still rankled. Even after all these years—and especially after the Beach takeover disaster—he felt that he (and Geosurveys) had been unfairly squeezed out of Santos. His integral role in establishing the company—virtually forcing them into exploring the Cooper Basin, for goodness' sake—and then negotiating the partnership with Delhi, had never been properly recognised, in his opinion. Rather, Bonython overstated his own contribution to Santos's scientific management in those early years, and avoided referring to Reg Sprigg at all. Shabby treatment indeed, in Reg's view. He was generally not a man to dwell on past grievances but writing his story had raked up a few coals of resentment. His spurned Ediacara fossils, his rejected doctoral thesis and his virtual dismissal from Santos still niggled.

'One of the greatest disappointments in life is to be let down, repudiated and scorned, by those whose careers and fortunes were due to you,' Oliphant soothed him. 'I know well from bitter experience, the strain and heartbreak of being spat at for achievement. But that seems to be the way of life for so many, who climb ruthlessly on the shoulders of others. I have come to terms with all this as a result of the deep enjoyment, and mental rest, given me by you and yours, in Arkaroola. I would do anything, commit murder if necessary, to help you do likewise.'

Oliphant gently reminded Reg that it was not worth upsetting himself over perceived injustices in the past. 'You are clearly far from well and should not be confronting now, after all that you have achieved nationally as a geologist and at Arkaroola as creator of a wildlife sanctuary, the … "bows and arrows of outrageous fortune".'

'You have built—in every sense of the word—magnificently. Let the edifice justify your every act. Just sit tight in beautiful Arkaroola, the wondrous beauty of which has captivated me, heart and mind, as it has so many others. You still have much to write, much wisdom to impart, much to enjoy of what you yourself have created.'

It was a loving tribute from his friend and Reg recognised the wisdom of his advice. Oliphant had sacrificed his own research career for the establishment of the Australian National University and the National Academy of Sciences, and then found himself strangled by bureaucracy and derided for his efforts. 'The White Oliphant' they called the particle accelerator he devoted more than a decade of his life to build in his Department of Particle Physics. Despite his key role in the Manhattan project, he had been politically smeared during the Cold War Senator McCarthy witch-hunts of the early 1950s; his American travel visa had been blocked and even his loyalty to his country had been unfairly questioned.

If he could rise above it all with dignity, so could Reg.

Reg also knew that Oliphant's comment about the state of his health was true. Much as he preferred to belittle it away, he knew that he was living on borrowed time. High blood pressure and severe angina pains were too-frequent reminders of his weakening grip on this mortal coil. By rights he should have been dead long ago. The devil looks after his own, he liked to say, when explaining away a cerebral haemorrhage, a quadruple bypass followed by a heart attack, acute hepatitis, several emergency landings in light aircraft and a couple of scuba diving mishaps that could have ended quite differently. But there was so much more yet to do, he sighed. So much to experience; so much knowledge to acquire. If only he had another lifetime in which to fit it all in.

ALTHOUGH REG HARBOURED some disappointment about the way his relationship with Santos had petered out, he never begrudged the company its success and he liked to keep abreast of their progress. He couldn't help but feel a small personal sense of pride at each new producing well in the Cooper Basin. The acreage in the Great Artesian Basin—now known more specifically as the Cooper Basin and the Eromanga Basin, which overlies the Cooper in some of the same areas—was now recognised as Australia's largest onshore hydrocarbon province. Natural gas reserves across the region were estimated at 100 billion cubic metres, plus liquids amounting to some 400 million barrels. There is no doubt that the oil and gas potential of the area would have been discovered eventually, but it was nice to know that the initial development was due to his own intuition and insight.

Like the majority of oil explorers around the country, Santos was enjoying a new era of productivity following a revival of drilling programs after the industry-wide malaise of the 1970s. Santos had particular reason to feel pleased. Late in the 1970s the company had made an exciting oil discovery at a well called Dullingari. Instead of the gas they had largely come to expect from the Cooper Basin holes, oil flowed out of the ground. The surprising thing was that the oil was coming from Jurassic-age rocks, much younger than the Permian rocks usually targeted. In the past they had always drilled through these shallower sediments as quickly as possible, anxious to get to the deeper, older rocks where they knew the gas—and some oil—to be. When the Dullingari oil came to light the realisation was instant: who knew how many other wells had been drilled through these oil-forming reservoirs without the operators recognising it.

In fact, there was a rumour that the discovery of the Dullingari oil really was a complete fluke. A road contractor was said to be flying over the Dullingari field, noting the gas wells all ready for production to begin. He also noticed a lot of contaminated oil on the ground. When he arrived at Santos's Moomba field he asked when he would be given the job of, 'cleaning up all that oil crap on the ground at Dulingari.' What oil!? came the reply.

Following confirmation at Dullingari, other old drill cores were checked and fields revisited. Jurassic oil was found above the Merrimelia and Moorari oil and gas reservoirs that had first been drilled more than 10 years earlier, and at new prospects called McKinlay and Marabooka. It was akin to discovering an entirely new oil province but instead of being many miles away it was separated only by vertical distance: Metres of rock representing millions of years between the end of the Permian time period when reptiles and conifers covered the earth, and the start of the Jurassic time period, 50 million years later, when dinosaurs and flowering plants abounded.

Late in 1981, the Jackson well in Queensland was drilled and oil poured out from multiple zones at a rate of 3878 barrels per day. It was the discovery that clinched the commercial feasibility of Cooper/Eromanga Basin oil.

By the time Santos's founding Chairman, John Langdon Bonython, passed away in April 1992 at age 87, Cooper/Eromanga Basin natural gas was supplying both Sydney and Adelaide; liquid hydrocarbons were flowing down a 659-kilometre-long pipeline to Port Bonython near Whyalla in South Australia and being turned into liquid petroleum gas for shipment to Japan; Jackson oil was being piped to Brisbane's port and refining facility; and Santos was a large international company with expanding interests in the Timor Sea and other areas. Bonython, too, had had a lot to be proud of.

WHILE SANTOS WAS veritably motoring along, Reg's other paternal oil interest, Beach, was stuck in neutral while court action to resolve the disastrous IRL directorship proceeded. The litigation dominated company affairs and hampered efforts by Beach to resume its core activity of searching for oil and gas. A regime of 'care and maintenance' was adopted by the new Board with careful husbanding of resources. Thankfully the Bodalla oil field had not been flogged off by the IRL raiders and production at the field brought in some much needed income.

The trial against Fuller, Main and Cummings; alleged IRL ringleader, Malcolm Johnson; and a host of IRL-related companies commenced in Adelaide on behalf of Beach and Claremont on the 6 July 1992. It lasted for ten months and was touted as one of Australia's biggest civil fraud cases. Every day there was a representative from Beach in court to monitor proceedings. Jean Shaw, the loyal Beach shareholder, was often there too, sitting with her knitting in the back of the court room.

'I was absolutely in awe of Fuller as a lawyer. He was brilliant!' she says anomalously. 'He came in his "gardening clothes"—shoes with soles half off that he'd picked up on his way in—and say he couldn't afford legal representation: He had to represent himself. He'd say, "On behalf of my client, Mr Fuller…" and gesture to an empty chair beside him. It was great theatre, and you didn't even have to pay to get in! But really I went because I wanted to see Fuller—who'd told a swag of lies at the last AGM—brought down.'

From her position at the rear of the room, Jean glowered at the fraudsters—including Beach's Reg Nelson in her disapproving gaze—until she realised he was not one of the enemies. Reg Nelson had been appointed Executive Director of Beach about a month before the trial started. He was a solid, confident presence who came to the company with impressive credentials and from the prior position of Director of Mineral Development for the South Australian Government. It was his job to report to Beach in Sydney on the daily court developments in Adelaide. He'd really only taken the position because Reg Sprigg had asked him to.

'You ought to get involved and fix this,' Reg Sprigg had urged when the evisceration of Beach had first come to light. Reg Nelson greatly respected the elder statesman of the petroleum industry and had always had a soft spot for Beach. He envisioned that the court action would occupy him for a couple of months and then he could get on with the real business of rebuilding the company. He never imagined that the legal morass would ensnare him for a year and that he would learn so much about forensic accounting and laundered money winging its way around the world in the space of 24 hours.

As the court case unfolded and the extent of the fraud against Beach was revealed, Jean became concerned for the personal safety of Reg Nelson and his fellow director Bob Kennedy, who was also deeply involved as an expert witness for Beach. There were rumours of people being shot in Swiss hotels; of having their cars tampered with. Who knew the limits of this murky world they were exploring.

Reg Nelson, too, felt apprehensive at times—especially when told by two of the key witnesses that he had good reason to be. But he stayed the course. Each Friday as he left the Adelaide Federal Court building he'd walk up Pirie Street to Arkaroola's city office and update Reg, or any of the Sprigg family who was there, on the week's proceedings in court. They all knew that, for Beach, this was a battle for survival.

On a personal level, there was another thing that happened in May 1992, that brought Reg much joy. It was an unexpected letter that arrived at Arkaroola from Sweden.

'I hope it won't be too much of a shock for you to be confronted with a voice from 45 years or more ago,' the letter began. 'Your name has cropped up now and again in the Swedish and British press particularly with regard to your famous "soft" fossil finds in the Flinders Ranges, but also with regard to oil. However, I hadn't realised the great extent of your achievements until I happened to see a new copy of the Australian *Who's Who*, in the central library in Stockholm with half a column of exciting enterprises and honours. I got the impulse to write and congratulate you…If you feel inclined to acknowledge my letter I'd be delighted, of course,

Sincerely,

Pat Een'

It was a letter from his ex-wife—the former Patricia Day—who had fled back to England after her affair with Teichert, less than five years after they were married.

Poor, unhappy Pat, Reg mused. She had been such a tortured soul when she'd sailed from Adelaide in February 1948. It had all happened so long ago that, thinking about it now, he felt like he was recalling a different person in a different lifetime. His life, since then had been so full; so fulfilling. He had certainly not harboured ill feelings against her over all those years. He only hoped that she had found the same happiness in her own life. It was wonderful, now, that she had gotten in touch.

Pat was now 69 years old. She had two daughters, two grandchildren, and a Swedish husband whom she had married 25 years ago. She had never been back to Australia, she wrote, but did admit to feeling nostalgic at times for, 'the outback, the distances, the galahs and cockatoos, and the starlit night skies.'

She had always loved the outdoors, Reg recalled. Sometimes she had been able to accompany him when he was out on field assignments with the Department of Mines. He remembered that she had been with him at Ediacara when he had discovered the very first fossil jellyfish: He hobbling about with his sore toe while she did the lion's share of making camp. He remembered, too, how much she had loved the misty blues of Mount Deception and the evening purple of the Aroona Range, and how they had both marvelled at the brilliance of an aurora australis display from sunset til four o'clock in the morning—the sky lit up in red and yellowy-green with yellow-white streamers dancing across the stage. They had spent the days walking the creekbeds studying animal tracks, collecting Aboriginal arrow heads on the drifting red sand dunes to the west and hunting for fossils. They had both been so young then, so carefree and idealistic.

Naturally, Reg replied to Pat's overture and the two began a monthly exchange of letters. They shared an interest in music, history and the natural world, and were never short of topics to discuss. He also told other of his correspondents how pleased he was at the renewed contact.

'It's so nice to be able to correspond so freely and without rancour after all these years,' he divulged. 'For me, it is a great pleasure to know that she is happy and her life is fulfilled.'

Griselda, too, approved of the contact. She knew she had nothing to fear from an ex-wife separated by 44 years and many thousands of miles. In fact, she and Reg began to speak of a visit to Sweden to meet Pat and her 'new' husband. They had so many overseas friends whom it would be lovely to see again. Reg joked that it would be their 'last' trip to Europe, 'a nostalgic final sweep to see old friends…' They tentatively planned the holiday for the winter of 1994.

It was certainly as well not to postpone things too long, Reg thought to himself as he looked back through his Christmas card list for 1992. Many on the list were wives who had lost their husbands throughout the year. Obituaries of old friends were now also the order of the day in most of the scientific and technical journals that he still received. And now the latest disturbing news was that his old friend and former colleague, Bruce Wilson, had cancer.

When he thought back over all his various associates in the field, there was none that he held in higher esteem than Bruce. He had been a loyal friend and quite the best geologist Reg had had the good fortune to work with.

While employed by Geosurveys in March 1966, Bruce and fellow Geos geologist Tony Cooney had discovered two pieces of a giant meteorite out on the Nullabor plain of Western Australia, 100 kilometres from Eucla near the South Australian border. They had been using the larger, 2 metre-high black object on the horizon—a rare point of vertical relief on the sparse flat plain—to take bearings for their geological survey. They eventually approached it to find that their compass-reference marker was a rock from outer space. The smaller fragment weighed more than 5 tons (5461 kilograms); the larger, almost 12 tons (11 786 kilograms). It was, by far, the largest meteorite discovery ever made in Australia.

The 11-ton fragment was sent immediately to the Western Australian Museum in Perth. Bruce had so much trouble finding an appropriate institution willing to section the smaller fragment for scientific research, that he almost rued the day he found it. Eventually it was shipped to the Max Planck Institute for Nuclear Physics in Germany, who agreed to distribute slices to museums and academies in England, America and Russia. To cut the iron behemoth they resorted to the method the ancient Romans had used to cut marble plates, using a rotating wire and abrasives. It took 180 hours with an iron-wire saw and carborundum to cut the first 200-kilogram slice.

It was named the Mundrabilla Meteorite after the railway siding on the Transcontinental Railway nearby to where it was found. Reg was pleased that the 2500-kilogram remainder of the smaller fragment, which now stands at the entrance to the South Australian Museum, would remain as a permanent tribute to his mate Bruce, no matter what the outcome of his illness.

Fortunately, one who remained in good form was Reg's dear friend, 91-year-old Sir Mark. When Reg sent Christmas greetings to Oliphant that year he wrote resignedly of the continuing dry weather. For weeks heavy dark clouds had been amassing with plenty of teasing thunder and lightning, but the tropical rain clouds from the north refused to release their load. Across the southern slopes of the village hills the Silver Wattles were in full bloom, but Reg hated to see the bare, grey trunks of the River Red

Gums in the dry creekbeds pointing their naked branches into the sky. Even if the summer rain did come, it would likely be too late, he thought, to save some of these majestic trees.

Oliphant wrote in reply of the hot, humid weather he was enduring in Canberra, 'making it impossible to do anything outside without sweltering in a Bombay climate! What a contrast with our first Christmas in England!' he wrote, recalling when he had first left Australia in 1927 to begin postdoctoral studies in Ernest Rutherford's famed Cavendish Laboratory at Cambridge University.

'We were invited by John and Elizabeth Cockcroft to drive with them on Christmas Eve to share a family Christmas in Todmorden. We travelled 100 miles in an ancient, bull-nosed Morris, the celluloid side-screens of which had long since fractured or disappeared altogether. Loads of rugs did not keep legs and feet above freezing point. After about five hours of careful arctic driving we were deposited at Elizabeth's home, where roaring coal fires in all rooms so far as we could see, restored welcome warmth.

'Snow fell during the night, but not heavily enough to prevent us all walking to the Cockcroft's home, where John's widowed mother had prepared an enormous midday feast for 12 people, with enormous turkeys, hams, all the bits and pieces, including loads of roast potatoes, greeted us. Our hostess was most upset if anyone refused a second, or even a third great helping. There followed an enormous, spherical boiled pudding, with oodles of egg-laden custard.

'After some post-prandial activities, details of which I have forgotten, we returned to Elizabeth's home to sleep it off!...

'The drive back to Cambridge was a nightmare. We soon encountered thickening fog and normal driving became impossible. So, clad in everything I could pull round myself, I lay on the running-board, from which I could get glimpses of the curb, signalling right or left as we crawled along. At junctions and crossings I climbed the posts with a torch to read the signs. This went on for over two hours, when it cleared enough to drive cautiously and slowly. We reached Cambridge at about 4.00 am! We all recovered after a good sleep, with no ill-effects. What it is to be young!

There seems little left to write about except to wish you all a New Year filled with happiness and fulfillment.
Warm wishes,
Mark'

WHILE REG AND Griselda's planned trip overseas was an event to look forward to, Reg still had plenty of activities at home to keep him occupied throughout 1993. There were two projects, in particular, that he had been ruminating on for a long time and continued to mull over.

One was a typical Sprigg departure from the expected; another grand sweeping entrance into what seemed to be an entirely new realm. He dreamt of a road tunnel through the Mount Lofty Ranges that formed Adelaide's immediate eastern boundary and separated the city from the population centres of the eastern states. Every truck, bus or car travelling from Adelaide to Melbourne, Sydney or beyond had to negotiate the sharp, inclined angle of the Devil's Elbow and other such hairpins to make their way from the city by the sea to the Murray River plains on the other side of the range. Noxious fumes belched out in black clouds as heavy vehicles laboured up the hill and tragic accidents claimed lives. A tunnel would provide direct access from Glen Osmond to Hahndorf without the time-consuming and treacherous ascent to Mount Lofty.

As Reg remembered it, it was an idea first proposed back in the 1930s by a senior engineer with the Engineering and Water Supply Department. It was then resurrected in Premier Tom Playford's day but Tom had 'knocked the project on the head' then, as being 'out of the State's financial reach.' The fact that it had been brought up at the time by the opposition leader would not have appealed to Playford either, Reg suspected.

Now, however, was a good time to revive the idea. It could be a great way to mark the State's celebration of the year 2000 and serve as a tribute to Playford at the same time, Reg enthused, in outlining his plan to Norton Jackson, the chairman of the South Australian chapter of the Australian Academy of Technological Sciences and Engineering, of which Reg was a fellow.

He prepared maps and geological plans of the Adelaide Hills to highlight the best site, and canvassed the idea to influential friends and scientific societies across the state.

'I feel I should seek co-authorship of some specialist in the area of tunnelling,' he acknowledged to Jackson. 'It should be someone with a *wide* view on such matters and who is not afraid to stick his neck out jointly with me. It *has* to be a "big thinker."'

The other concept that engaged Reg's mind was closer to home—intellectually and geographically. He thought long and deeply about Mount Gee. It's preservation had been one of the main motivators for the purchase of Arkaroola in the first place and he had never lost his fascination with its craggy red presence. The formation of the 250-metre-high mountain seemed to be extremely complicated and as successive mining companies had probed it for uranium its geological mysteries had been heightened rather than lessened.

For many years Reg had been angling for an opening into Mount Gee. He wanted a small adit—nothing big or obtrusive—to see inside this crystal mountain. The mining companies spoke of large subterranean voids where drill bits had fallen 20 metres through open space and been lost. Reg imagined great hidden caverns lined with beautifully formed crystals of clear quartz, purple amethyst and pretty green fluorite. Imagine being able to take students and tourists actually inside the mountain to see the inner workings of this ancient hot spring. It would be a truly unique experience.

He had agitated the Department of Mines over the years for permission to tunnel in, but the bureaucracy had always refused, even though Reg offered to pay for it himself and access would be strictly controlled for educational and illustrative purposes. Mineral exploration companies could pepper their drill holes all around, it seemed, but Reg's concept of a walk-in passage into the mountain appeared to be too foreign to even consider. Never mind; he re-offered the idea every now and again.

The other intriguing thing about Mount Gee was the 'breathing' air vent that he had discovered. One of the drill holes bored by CRA Exploration must have breached one of the internal cavities that Reg was so intrigued by.

This one obviously opened to the outside at some point because air moved up and down, in and out, through the hole with quite some force. It was enough to blow a hat up in the air if it were placed over the 10-centimetre opening. Holding you hand over the hole you could feel the warm moist 'breath' of the mountain rushing up into your palm.

Reg had spent much time investigating the phenomenon and had learnt that the flow of air was related to atmospheric pressure. After two weeks of sitting up all night taking hourly barometric readings he had confirmed that the mountain 'breathed' with the weather. About 36 hours after a low weather system passed over, the mountain 'exhaled'. When a 'high' passed over, the mountain 'inhaled.' Just one more thing that made this mountain so special.

While Reg continued to 'chew over' the geological history of Mount Gee, he also turned his thoughts back in time to the ancient beach system off the south-east coast of South Australia. A cray-fishing contact in Mount Gambier had enquired if Reg would be interested in a voyage out from Robe and Beachport to trace the sea bottom with echo sounders and fathometers. He knew that Reg was supremely interested in the system of fossilised sand dunes that marched inland in parallel many miles from the present coastline. Reg believed that the same pattern of former coastlines continued out to sea and he was grateful for another chance to 'see' the submerged dune system for himself.

He had first written about this system of stranded sea beaches back in 1948 when he had presented the results of his research at the 18th International Geological Congress in London. He had correlated the remarkable set of ancient beaches off the south-east coast of Australia with periods of high and low sea level associated with the growth and retreat of glaciers. He postulated this was all tied up with the so-called Milankovitch cycles which predicted the amount of sunlight falling on the earth in any one spot according to the varying elliptical path of the earth's orbit around the sun and the changing tilt of the earth upon its own axis.

He had been complemented at the meeting on the quality of his evidence, although some people thought the fit of Reg's field data to the

ONE LIFETIME IS NOT ENOUGH

theory was *too* good and cast some suspicion on the whole idea. When *Saori* had been afloat, Reg had used the opportunity to make observations of the ancient dunes that progressed seaward from the modern coastline—now-submerged beaches that had been high and dry during the last Ice Age. He was more convinced than ever that the multiple oscillations in sea level correlated perfectly with the Milankovitch climate cycles. He believed that these sand dunes—now way out to sea—were formed when glaciers sucked up the oceans and sea levels were low. He felt certain that aborigines would have populated these old dune systems and that one day their shell middens would be found atop the submerged sand dunes. He failed to understand the apparent lack of interest in his findings. It remained one of his many passions, nevertheless.

So when, early in 1993, he was invited aboard his friend's boat he didn't squander the opportunity. In the two days they spent traversing the continental shelf out from Beachport, Reg found clear evidence of three major submerged beaches from the last Ice Age. The beaches lay parallel with the modern coast at 20 to 25 fathoms (37-46 metres), 40 to 45 fathoms (73-82 metres) and 50 to 60 fathoms (92-110 metres). He named them Geltwood, after the site of Beach Petroleum's first deep oil well; Admella, after the ship that had foundered on Carpenter Rocks nearby and from which his great, great uncle had rescued the last remaining survivors; and Waterwitch, after the first seagoing vessel to navigate the mouth of the Murray River and which his great grandfather and great, great uncle had both piloted.

Reg was profoundly satisfied with the results of the expedition.

'The offshore operation was a resounding success and I am still in the process of reducing the data and glowing over its perfection,' he wrote in gratitude to his friend in the South East. At 74 years of age it was wonderful to feel that he was still making a scientific contribution.

THE YEAR 1994 was shaping up to be a good one. Reg and Griselda's overseas holiday was planned for May/June. Before then they had a crowd of ex-Geosurveys members coming to Arkaroola for a reunion in March, and there was also a reunion of ex-Radium Hill residents planned for the Easter weekend in April. Life was as busy as ever.

The tight-knit Sprigg family, too, had expanded. In November, Douglas had fathered a son, named Mark, in honour of their great family friend. Oliphant was moved by the tribute and doted on his namesake upon meeting him at Arkaroola. Two hounds also joined the clan—a pair of Anatolyan Shepherds name Raki and Jezebel. 'They are like lion and lioness the two of them—magnificent creatures that will grow to the size of horses!' Reg described them to a friend. 'They are quite cuddly and great fun but excessively powerful…'

Unfortunately both Reg and Griselda were to bear the full brunt of their colossal strength when the two dogs bounded down Margaret's driveway to greet them in Adelaide and knocked the two Spriggs flat on their backs. Both were confined to bed with bruised backs and legs. Then—two days later—Reg awoke to find himself paralysed down the right side. He couldn't move his arm or his leg and his speech was all slurred. Minor stroke, the doctor diagnosed; and thus their holiday plans, too, were knocked flat. Two weeks bed rest was the immediate prescription, with all thought of lugging heavy baggage around the world curtailed until spring at the earliest.

The holiday postponement was disappointing but just then, Reg had more pressing concerns. One hundred Geosurveys colleagues were travelling to Arkaroola in a couple of weeks' time to celebrate the 40th anniversary of the founding of the company and Reg had promised a commemorative time capsule. He had planned a rock and cement cairn atop Observatory Hill and he was determined that it would be ready in time. Bed rest be buggered, there was too much to do.

After one day recuperating he was back in the yoke. Staff problems at Arkaroola were looming, with the disgruntled cook threatening to sabotage the up-coming reunion weekend. Reg hired a second cook just in case. Their motel supervisor had just given a month's notice, the bore water supply was faltering, electrical wiring and the roofing on the motel needed replacing, and the village sewerage system was up for renewal. So much money to find and so much work to do. All that and the cairn to complete

ONE LIFETIME IS NOT ENOUGH

as well. It was a lot to contemplate for a young man in full fitness. For a man of 75 still troubled by a sore back and unsteady on his right leg it was surely too much.

Reg admitted to his friends that the labouring jobs, in particular, were taking their toll. 'Slowly I am recovering but the concreting etc that I have been doing in relation to the Geos reunion and Sir Mark Oliphant's visit this weekend has helped me into repeated minor "relapses",' he revealed. 'I fell over uncontrollably six times yesterday while doing the concreting to build the cairn to house the time capsule. The boys were dragging their feet—"too hot," they said. Still, it is now 90% complete and can be finished in time.'

Ex-wife Pat, over in Sweden, was appalled when she heard what he'd been up to.

'Although I no longer have the "right" to scold you, I must say that it's absolutely wrong to literally risk you life in this way. The wretched cairn could well have been *your* memorial! It's crazy to ignore the very *grave* warning (no bad joke intended) that your luckily slight stroke was. Now, I won't risk our friendship by any more finger-wagging but just say that we are looking forward very much to meeting you both—so just see that you make it!! Please…'

Griselda—with Reg at Arkaroola—was too busy herself, to see that Reg laid off the labouring and took care of himself. She was working in the motel from eight in the morning to nine or ten at night most days. Anything to keep labour costs down.

'At our age it's unreal,' Reg agreed, when asked why he and Griselda didn't begin to wind down and retire, like the rest of their friends. He did worry about Griselda's work load, though. She seemed never to stop. They were as bad as each other.

When he spoke to his doctor about the lingering pain in his leg, back and neck a month later, he received little sympathy:

'What else can you expect?' the doctor asked. 'You've never rested, and have a guilty conscience and played-out body. Need I say more.'

Reg knew he was right.

301

Still, despite, their ailments they enjoyed the Geosurveys reunion immensely. The cairn was finished in time and the weekend came together perfectly, as is often the way. It was a wonderful three days of reminiscing and story-telling. Reg was amazed at the camaraderie that still linked his old Geosurveys crew together. It was now 40 years since the company began and almost that long, too, since many of these men and women had worked together, or in some cases, even seen each other. They all spoke so fondly of Geos and how it had marked a special time in their lives. They had really felt like pioneers, back then—venturing into the desert; under the sea; places where truly no one had been before. And at the centre of it all was Reg—this man who had motivated and inspired them all, and made them feel that they were a part of something special. Their presence here now, was a testament to him.

By April of 1994, Reg had almost fully recovered from the fall *and* the stroke, and the overseas holiday was back on the agenda. They would go in September and enjoy the Northern Hemisphere autumn.

The first stop on their holiday would be a visit to the Eens—Pat and her husband, Gillis—in Stockholm. From there they would travel to friends in Germany and Switzerland, then over to the United Kingdom where they would holiday in Jersey for a week, then make their way from Devon to the Scottish border, visiting friends along the way, and finally to Griselda's sister in Glasgow. They would return to Australia in early November and have the quiet summer season at Arkaroola to recover from their exertions.

Over in Sweden, Pat worried that Reg might never make it, what with the way he drove himself so relentlessly. But when they arrived at the airport on that Thursday in mid-September, there were Reg and Griselda, happy to see them. It was a joyful reunion for the two former young newlyweds and an endearing introduction for each of their spouses.

For Reg, catching up with Pat and her family was a real highlight—'a late-life plus,' he pronounced it. He and Griselda were wined and dined magnificently but there was also an emotional unburdening that Reg had not expected. One night, after dinner, Pat announced that she had something important to say—a confession of sorts. She then set about describing how her life had unfolded after she had abandoned Reg and sailed to London in shame 46 years ago. It was a sad tale that finally ended happily with her meeting and marrying her current husband.

Reg and Griselda were deeply touched. It was clearly a story that Pat had wanted to share with the other three and Reg admired her courage in being so honest. Apparently her repudiation of Reg had weighed on her mind all those years.

'I was completely humbled by the obviously suppressed emotion of it all,' Reg admitted to Oliphant when he wrote to tell him of the episode. 'Sorry to load this on you Mark, but it was an unusual experience and does help to close off what was, at the time, a very unpleasant experience. Griselda, as always, took it like a rock.

'All I can add is that I have shared much of my life with two remarkable women. Some people are lucky—and I am certainly one.'

FROM EUROPE IT was over to London, where Reg's itinerary stated, three days 'at leisure.'

'Like hell,' Reg faxed daughter Margaret in mock reproval. 'Some young lady, Marg by name, loaned my itinerary to Reg Nelson. Now I am to be met at the Ibis Hotel, Heathrow, by high-powered lawyers to be used in an action by Beach Petroleum (as a private shareholder) to claim $15-20 million off the Fuller-Johnson mob and the take-over company. What's in it for me? Nothing. The same old story. If only I was a lawyer on $1000 to $4000 per day "appearance" money.'

Margaret was apologetic for intruding on her parents' holiday, but she also knew how aggrieved her father had been over the whole takeover action and the subsequent financial swindle. If there was any way Reg could help in recouping some of the losses, she was sure he would want to know. And Reg Nelson had been so insistent that Reg Sprigg's contribution was vital. Her father deeply admired Beach's new Executive Director and the way he was carefully nurturing the company back to good health. Marg knew that he would want to help.

Upon arriving at his hotel there were letters awaiting him from Beach's London solicitors. Reg worked long into the night recalling everything he could concerning the fraudulent actions of IRL in the takeover of Beach and the subsequent purchase of the depleted Burbank oil fields.

The ten-month trial of Beach versus the IRL group had concluded in June the year before (1993), with the Federal Court judge awarding Beach damages of $44.45 million, plus costs. The purported liability of $6 million for part of the loan to Beach to buy the Burbank fields was extinguished in the same judgement. The Court found that there was a conspiracy to defraud Beach by its former directors and a conspiracy to mislead and deceive Beach in relation to the Burbank transaction. The IRL directors of Beach had caused the company to purchase the Burbank oilfields at a grossly inflated price—$32.25 million more than it was worth or almost 10 times what had been paid for it less than a year earlier. The deal had been done in disregard of valuation advice, and behind a veil of severe conflicts of interest.

Overall, the court verdict was a good outcome for Beach and the new Board of Directors saw the result as a vindication of the company's battle for justice as well as an assurance that the company could continue to exist and begin to return to profitability.

Reg Sprigg's current role in the ongoing war was to legally declare that, as a private shareholder, he would have roundly rejected the purchase of the Burbank, Oklahoma, oilfields if the proposition had been put to the shareholders—as it should have been. The IRL directors broke Australian Stock Exchange rules by never revealing that they stood to personally gain from the deal. They were operating on both sides of the transaction. Reg Sprigg was happy to make it perfectly clear that his personal right to veto had been stolen from him, along with the millions of dollars of Beach's money.

AFTER THREE DAYS in London, Reg and Griselda were on the train to Portsmouth and the ferry to Jersey, and the bitter after-taste of the Beach fraud action was behind them once again. They spent a week in marvellous island seclusion and then two weeks motoring up to Scotland staying with friends along the way. It was on the 30 October that their holiday took a sudden turn for the worse.

It was a black Sunday night and driving in a pelting rain storm on the M6 motorway was treacherous. Cars swept alongside them in 140 kilometre-per-hour waves of water. The windscreen wipers struggled to clear the view. Both Reg and Griselda were exhausted from the hectic pace of holidaying and the 500 kilometres they had driven that day. Griselda, especially, was wan and lethargic. All they wanted was to find their way to sister Margaret's house and collapse into bed. Trying to read exit signs in a high-speed wet blur was impossible.

When a large lorry cut in front of their car they were blinded in its massive wake. Griselda screamed and fainted. Reg somehow kept the car on the road. In a daze he found Margaret and solace for the night. The next two days passed in a fugue of disorientation…Griselda in intensive care…days of no memories…a stabbing pain in his chest…and then…Nothing.

WHEN REG SURFACED into consciousness, he was in the Royal Alexandra Hospital in Paisley. He was now a patient in the very same hospital in which Griselda had nursed. It was here that she had taken the phone call from Australia and agreed to become his wife, all those long years ago. Now look at the two of them—Griselda admitted down the hall with bronchial pneumonia; he in coronary intensive care, teetering at death's door for almost a week with a heart attack followed by two strokes. So this was the price of trying to pack too much into one's holiday, he thought groggily. What a sorry bed they had made for themselves to lie in.

While Griselda's condition began steadily to improve, Reg's situation was far more critical. He had almost died, the doctor admitted. Only the news that Margaret and Douglas were on their way over from Australia, pulled him back from the brink. It was then that his body rallied. And when they actually arrived at his bedside a couple of days later, his recuperation began in earnest.

The therapeutic effect of his children was quite remarkable, his doctor announced. Reg's heart was returning to normal size and his vital statistics were improving by the day. If progress continued at that rate they would soon be able to consider an angiogram to determine if Reg's obstructed blood vessels could be cleared or propped open, or if further bypasses were necessary.

But there was still a long way to go.

The 7 November—the date of Reg and Griselda's original scheduled departure from Britain—had come and long gone. He'd been barely conscious on the day that their flight departed. Two weeks later he was feeling better but far from up to the rigours of facing an international flight. Moving from the Royal Alexandra Hospital to the Ross Hall Hospital in Glasgow was one thing. Getting himself back home was quite another. The health insurance company even forbade him to fly until his condition was much more stable. He began to despair at even being home by Christmas. But at least he knew all was well at Arkaroola. They had even had some rain—6.6 millimetres. Every little bit helped.

The flood of letters and phone calls from Australia warmed his poor damaged heart and he even began writing some replies. It was what he had always done; and putting his thoughts down on paper made life seem a little more normal.

'Great to hear of rain at Ark,' he wrote to the manager at Arkaroola. 'How's the chook house and rock wall? Would love to give you the traditional big hug,

Much love, Reg'

He also began writing some notes to himself—a diary of sorts to record the phone calls and visitors that coloured those short, dull days. He found it more difficult to stay cheerful as the northern winter closed in. The sun seemed hardly to rise at all, its feeble rays struggling to penetrate the greyness outside. The prospect of spending weeks confined to this sterile room was a dreary one, despite the efforts of the bright young nurses and their gay banter.

When the doctor pronounced on the first day of December, that Reg was ready to face the angiogram, Reg, too, felt ready to advance. He couldn't just lie here indefinitely, in a cold foreign land, far from the rocks and the red

Australian earth that he loved. The doctor had explained the procedure and carefully explained, too, that because of his age and his infirmity it was a much more dangerous intervention for Reg than it was for others.

'One in 50 risk of kaput,' Reg noted to himself, instead of the usual risk factor of one in 1000. Still, 'will go ahead with angiogram,' he noted. 'Signed permission.'

That was Thursday night.

On Friday morning he wrote to his dear friend Oliphant:

'Dear Mark,

By the time you get this there is a remote possibility that I will be undergoing Gaian recycling. Shortly I am going in for an angiogram. Doc tells me the chances of hopping the twig in the process is down to 1 in 50. So here's hoping…I go in in a few hours.

Mark, just in case something goes wrong. May I say thanks (what an inadequate word) for all you have been and meant to me…

Hope to see you at Arkaroola at about beginning of March.

Deepest affection,

Your Reg'

It was the last letter he ever wrote.

Inaria karli

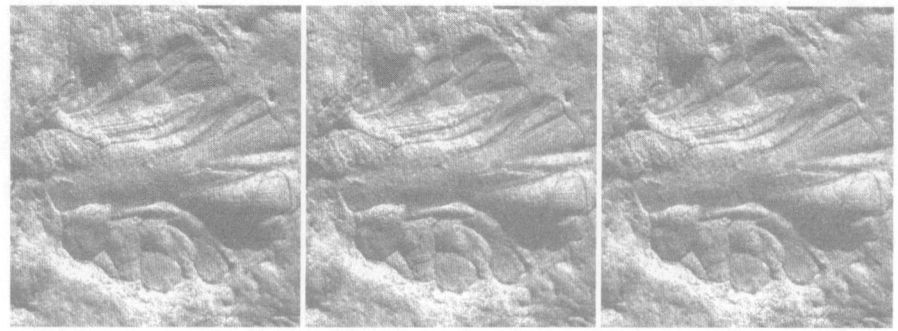

CHAPTER 14

THE SPRIGG LEGACY

REG SPRIGG DIED in Ross Hall Hospital on Friday, 2 December 1994.

Despite the gravity of his illness no one ever really thought that he would die. He had always pulled through before. He seemed so invincible. How on earth would they get by without him?

For Griselda, it was the day the light went out in her world. She would return to Arkaroola and gaze upon the Mawson country that her husband had loved, but there was no Reg to bring it all to life. Without him the hills lost their ruby glow, the crystals failed to sparkle and the sun-drenched granites looked tired and dusty. It was how she felt, too: Used-up; freeze-dried; desiccated. Dead.

She couldn't bear to open the thick curtains covering the full-length windows in their lounge-room. The view was too painful. Her despair too savage.

Everyone tried to keep her busy, as if that would make her forget the magnitude of her loss. So, she was busy. But still bereft. The only thing that kept any of them going was the knowledge that that was what Reg would want. It was now more important than ever that Arkaroola survive. It was the culmination of his life's work and now stood in tribute to the man they had lost.

Four days after Christmas the family climbed into the plane and Doug flew them out over Mount Painter. It was exactly the kind of flying day Reg had always loved—beautiful, clear blue skies with a bit of a cross wind providing bumps for added interest. Griselda found it harrowing, but she was there at Reg's request: 'Just put me in an orange box, take me to the local incinerator and then scatter me over Mount Painter,' he had instructed her years before. She only wished she had never had to follow his instructions through.

SINCE REG HAD died in Scotland and been cremated shortly afterwards at a small funeral service in Griselda's hometown of Paisley, there had been no opportunity for friends and family in Australia to pay their respects. The media had marked the passing of Australia's 'most colourful, enterprising, entrepreneurial, flamboyant and successful geologist' with a series of obituaries in newspapers and magazines:

'A poll of his geological colleagues would undoubtedly rank Sprigg as the person to make the greatest individual impact on Australian geological knowledge since the 1930s, if not ever,' wrote *The Australian*. 'It is unlikely we will see again such an extraordinary combination of scientific acumen linked with an intense love of the Australian environment and the urge and ability to communicate at all levels.'

The chance for those outside the media to add their voices to the tributes came early in February at a memorial service at the University of Adelaide. More than 500 people joined South Australia's Governor, Dame Roma Mitchell; the university's chancellor; and other distinguished guests in Elder Hall to hear past colleagues of Reg's speak of his contributions to the fields of geology, conservation and tourism. The final speaker was South

Australia's 93-year-old former Governor, Sir Mark Oliphant. He described how he had first met a then 28-year-old Reg at Mount Painter, back in 1947:

'I liked all that I saw of that young man...Then in 1971 I was honoured to become the Governor of this state and received an invitation shortly afterwards to visit a new privately established sanctuary—Arkaroola—in the northern Flinders Ranges...It was a wonderful experience to meet again the person who had so attracted me at Mount Painter—on the summit of which his ashes now lie. There began a very special friendship—the greatest of my life. We exchanged letters every week, never exhausting subjects for discussion.

'Reg was able to awaken in the hearts of many the magic of Arkaroola. Even transient visitors experienced uplift from the environment, leaving memories of special significance. Seated with a companion overlooking a deep, steeply sided valley, watching the crimson light of the setting sun creep slowly up the opposite cliff, I had the most remarkable, the most enduring experience of my life. A wondrous gift from Arkaroola.

'I can imagine no more enduring, no more fitting memorial to Reg, than Arkaroola. In the past, Griselda and their children have upheld Reg in very practical ways in his pursuit of the vision of Arkaroola. They will I am sure, continue to follow that star, bestowing on visitors, that blessing which has been mine.'

IN THE YEARS that followed, tributes continued to accumulate and recognition of some of his achievements continued to grow.

The Australian Petroleum Production and Exploration Association changed the name of their Gold Medal—awarded annually to recognise an individual who has given outstanding service in promoting the objectives of the Australian oil and gas exploration and production industry—to the Reg Sprigg Medal, to commemorate their founding chairman. Reg often liked to recount how he only gained the position because he arrived late to the first meeting in October 1959 and the only seat remaining was that of the Chairman's position at the head of the table. It belied his six-year reign as Chairman during the birthing years of the industry, and his role of councillor for the 20 years following.

Griselda presented the first APPEA Reg Sprigg medal in April 1995.

In June 1996 the Exploration and Development Manager for Santos wrote to tell Griselda that they had named a wildcat oil exploration well after Reg. The Reg Sprigg-1 well was spudded on the 25 May 1996. Within a month it was spouting 140 barrels of oil per day from the Jurassic-aged Hutton Sandstone at 1930 metres, and 350 barrels per day from the older and deeper Triassic Tinchoo Formation at 2220 metres. A new oil field had been discovered.

Reg Sprigg-1 was located 115 kilometres north-east of the Moomba gas plant—30 kilometres north of the Innamincka-1 well that was spudded in 1959. It was right in the heart of the original Santos acreage that Reg had procured for them back in 1955. How he would have loved the ironic perfection of it all…The closing of the circle, as it were.

By June 2006—exactly 50 years after South Australia Northern Territory Oil Search finished drilling their very first oil well at Wilkatana—Santos was South Australia's top company, with 1600 staff and annual revenue of $2.8 billion. The company had drilled more than 1070 oil and gas wells in the Cooper Basin, generating more than $25.3 billion in revenue and funnelling royalties worth almost $1.5 billion into State Government coffers. It plans to drill up to another 1000 wells in the Cooper by 2010.

Beach, too, continued to derive success from its Great Artesian Basin acreage. It was a long, hard slog but Beach was slowly clawing its way back to prosperity. The only real asset it had managed to hang on to during the IRL era was its Bodalla Block oil fields in south-western Queensland. They continued to out-perform expectations and provided a steady income stream to fund the growth of the ravaged company.

Reg was never to know that his final deed for Beach—stepping forth to represent the thousands of individual Beach shareholders who had been screwed by IRL—would be instrumental in returning 10 million dollars to the company. It would be the bulk of the money ever retrieved. IRL directors Michael Fuller and Joseph Cummings were declared bankrupt and the 45 million dollars awarded Beach in the court settlement has never been fully recouped. The outstanding amount today equates to well over 100 million dollars.

"It can be said that [Reg Sprigg's] enthusiasm, tenacity and dedicated support has been a strong component in the company's survival and revitalisation," Reg Nelson's tribute to Reg read in Beach's Annual Report for 1995. "He was an active and vocal critic of actions taken by directors when Beach was swept into the vortex of the Independent Resources Group; in particular concerning the purchase of the Burbank oilfields in Oklahoma; and in this, he can be said to have served Beach continuously from its birth in 1962 until his death. He rejoiced to see the company survive intolerable injustice. Only a month prior to his death, he provided assistance in certain legal matters that may provide a valuable and tangible legacy to the company…"

Neither did Reg Sprigg have the opportunity to appreciate that not all of the lawyers involved in the court case against the IRL group were the "$1000-dollar-a-day" vultures he perceived them to be. In fact, Beach's legal team at law firm Piper Alderman, largely carried Beach's costs during the court proceedings, sustained only by their faith in Beach's position and the sporadic payments that Beach could afford. They were finally paid the balance owed them after the court settlement and continue to represent Beach to this day.

Under Reg Nelson's stewardship Beach slowly began to regain strength. In 1996 it was worth about $10 million. By 2003, it was approaching $100 million; and by September 2006 it was worth $827 million and debuted on the Standard and Poor's/Australian Stock Exchange's S&P/ASX 200 index, which lists Australia's top 200 public companies according to market capitalisation. At the same time, in another ironic twist, Beach acquired Delhi Petroleum by bettering Santos's $474 million takeover bid. The company Reg had founded when his Geosurveys had been spurned by Santos, was now the new owner of the rival company that had edged him out. He would have revelled in the humour of it all!

Beach's acquisition of Delhi handed it a 21% stake in more than 200 oil and gas fields in the Cooper and Eromanga Basins. Meanwhile they continued to make new discoveries in their own wells in the region.

Besides the Great Artesian Basin, Beach's other long term interest had been the south-east coast of Australia. The company maintained an exploration programme in the Otway Basin that had always been one of Reg's favourites. In 2005, they extended their acreage further east by buying a half interest in a gas and oil project in the offshore Gippsland Basin—better known as Bass Strait. Working with their joint venture partner, Anzon Australia, Beach installed a unique floating production, storage and off-take vessel and began selling oil from the Basker and Manta fields in January 2006, breaking the 40-year strangle-hold that the discoverers—BHP Billiton and Exxon Mobil—had had over Bass Strait.

The *Crystal Ocean* that Beach and Anzon hired was the first floating production facility in south-eastern Australia and enabled commercial development of a previously marginal oil field. The compact 101-metre-long vessel is connected to a turret mooring on the sea floor 155 metres below via flexible composite steel flow-lines. Forty thousand barrels of crude oil (6.3 million litres) can be processed and stabilised on board before being transferred to a large shuttle tanker for transport to a refinery or shipment further afield. The shuttle tanker *Basker Spirit* is one of the largest oil storage tankers to be permanently anchored in Australian waters, and is capable of storing 680 000 barrels (60 million litres). Beach was now playing with the big boys.

By 2007—thirteen years after Reg's death—Beach was enjoying its most prosperous year in its 45-year history. The company had the greatest revenue increase amongst South Australia's top 100 companies, lifting revenue 248% over the previous year to $567 million and placing it in tenth place overall. It was extending its exploration interests overseas once again and diversifying its Australian interests to include coal seam methane production and the pursuit of geothermal energy. This last, in particular would have earned a hearty seal of approval from Reg.

Record annual production of 9.4 million barrels of oil equivalent (boe) and proved and probable reserves of 90 million boe placed Beach within the top five petroleum companies in Australia. It was a long way from the IRL days, when Beach shares were listed on the Exchange at one cent each—if anyone could be found to buy them.

These days, Jean Shaw's biggest worry is that another company will try and 'head-hunt' Reg Nelson away. 'I'm scared stiff,' she says. 'He dragged the place up from nothing. He really is quite brilliant.'

She watches over Beach's progress like a protective sentry, always taking the time to personally meet any new director and never missing an Annual Meeting.

'If I'm planning anything out of Adelaide I ring up Beach and ask when the AGM is and plan around that. I wouldn't miss it for quids.'

She reads the Annual Report from cover to cover and phones Reg Nelson if there is anything she needs clarified. The shares she has held in the company, through good times and bad, for more than 40 years have earned her that much.

ON THE SCIENTIFIC side, too, Reg Sprigg's contributions have withstood the test of time.

He was predicting climate change, the need for alternative sources of energy and depletion of the world's oil reserves well before it was fashionable. He spent thousands of dollars designing and building wind turbines to generate energy, for which he was condemned for sight pollution and the machines sabotaged. For ten years he maintained and monitored a private wind station in the Simpson Desert and other places to determine the best locations for farming wind. How Doug hated shimmying up the swaying, spindly 10- to 30-metre-high poles every month to change the roll of paper in the recording devices.

Reg correlated prevailing winds through the centuries with formation of the Simpson Desert sand dunes and compared them to the ancient beach dunes of the South East. He embraced nuclear energy as the only viable solution to fill the energy gap before the hydrogen fusion energy that he and

Oliphant dreamt of, becomes available. He recognised the need for stringent environmental conservation regulations and drafted the APEA code of environmental practice that the petroleum exploration industry abides by today. He experimented with oyster, lobster and abalone farming because he knew that the burden on the world's oceans would increase with the burgeoning population and he feared that there wouldn't be enough food to feed us all. He worried about religious wars, political conflict, atomic annihilation and global pollution, and meanwhile fought to preserve a small patch of dry, hard ground in the Flinders because it deserved to be saved.

Today, ice cores from the Antarctic reaching back almost a million years show how temperatures, carbon dioxide levels, dust and methane concentrations vary in concert with the so-called Milankovitch cycles of natural climate change. Reg stood on a beach in Australia's south-east 60 years before and intuitively predicted the same thing. He correlated the remarkably preserved series of stranded beach dunes there with alternating periods of glacial and inter-glacial phases stretching back almost 600 000 years. He determined that each of the 17 beaches corresponded to a separate phase of high sea level which correlated with Milankovitch's tables of the maximum amount of summer sunlight falling on the earth. Contrary to common belief there were not four or five glacial periods, but more like 20, he proposed.

No one was ready to hear it then. What is accepted science today was completely ignored when Reg first presented it 60 years ago. His was the universal experience of those condemned to be ahead of their times. Over and over again, Reg pursued other scientific enigmas while waiting for the rest of the world to catch up.

'I have had the great satisfaction to have my previously published (1948) theory of climatic control of glaciation for the past million years or so (based on Milankovitch Radiation theory) accepted internationally,' Reg wrote to Oliphant in March 1978, 30 years after he had presented his findings in London and when the rest of the scientific world *was* beginning to catch up. 'Despite the ridicule down the intervening years it is now quite clear that glacial cycles within an Ice Age climax at about 90 000–100 000 year intervals. The literature is now being flooded with evidence to this effect also from sea floor coring and other directions including geomagnetic reversals. So, bad luck when one publishes unpopular ideas too far ahead of time!'

Marine geologist, Dr Chris von der Borch, believes that Reg's foresight into natural cycles of climate change is entirely under-appreciated.

'His scientific work was really pioneering, particularly the climatic stuff,' he asserts. 'As far as I know, that paper of Reg's was the first time it was ever said—the Milankovitch correlation—which means he should be famous. It's just amazing the way he put it together like that. He was right at the very cutting edge. After that, the Americans got into it—and I don't think they ever cited Reg's work, because it was in this obscure journal—and then everyone took it up and ran with it and now it's the accepted thing. It's really one of the basics of climate change that people look at to try and extrapolate from, to see what's happening now.'

Another area where Reg had to wait for science to catch up with him was the complete characterisation of his River Murray submarine canyons. In March 2006 a research team from the Australian National University (ANU) discovered the former route of the River Murray meandering across what was once an ancient coastal plain. The prehistoric river wended its way south-westerly across what is now called the Lacepede Shelf and tumbled into the sea via the Murray submarine canyons just as Reg had predicted.

'Reg Sprigg did some formidable work,' agrees Dr Patrick De Deckker, Professor of Earth and Marine Sciences and lead investigator of the ANU team. 'We basically updated his article of 1947. He was pretty good. He was postulating then…But he had a pretty good idea. He estimated sea level rise to be about 100 metres, which at that time was just amazing. But he did that. He had a lot of foresight.'

The Murray River submarine canyons that Reg defined as being at least 1400 metres deep—with the equipment available to him then—have now been found to extend down 5 kilometres or more and are at least 80 kilometres long. They are some of the most gigantic canyons on Earth, on a par with America's Grand Canyon. The deepest and largest of them is now named Sprigg Canyon, in his honour.

THE SPRIGG LEGACY

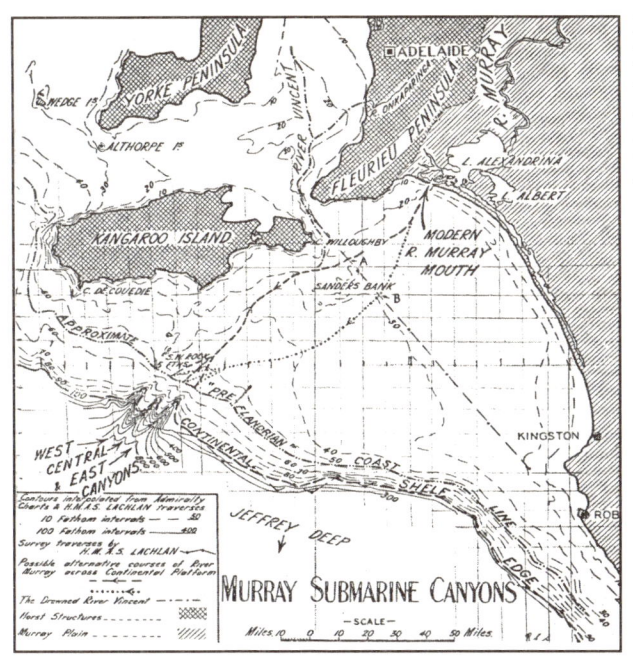

Reg's 1947 diagram of the Murray submarine canyons that he discovered and his two predicted routes of the River Murray across the continental platform.

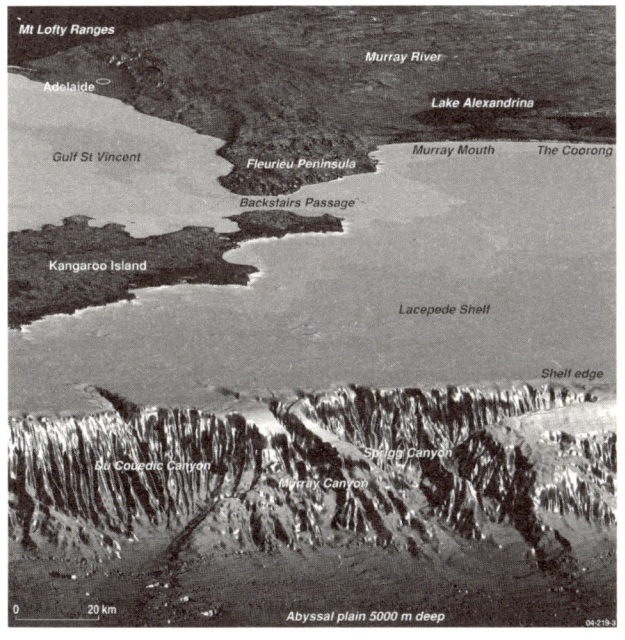

2005 satellite image of Sprigg Canyon, the largest of the Murray River submarine canyons that Reg discovered. Satellite imagery draped on onshore topography courtesy of Geosciences Australia.

'At that stage Reg was really pioneering a whole series of things—sea level change; climate change, which not many people had worked on; and also the submarine canyons off the Murray mouth,' says von der Borch, who named Sprigg Canyon. He collaborated with Reg in the early 1960s when he was the only scuba-diving marine geologist in Australia, just before Reg himself took up diving.

'I remember Reg going for his first snorkel. He had a very funny mask on—that he'd bought—which was a full face mask that had two snorkels and two ping pong balls in it, and they never worked very well. Then, of course, Reg went ahead with Darby von Sanden and did very deep dives.

I never actually scuba dived with Reg. His dives were too deep and dangerous!' Chris laughs. 'That's a pretty scary dive he did off Robe,' he says, referring to the dive Reg, Darby and Dave Burchell did to 82 metres using standard compressed air. It was believed to be an Australian depth record at the time.

'It may still be,' says Chris. 'It's incredibly deep; particularly with the old-fashioned gear. The modern gear, they've got mixed gas, they can do it. That's well beyond the danger zone for nitrogen narcosis. Amazing! You wouldn't catch me down there,' he laughs.

The research cruise that von der Borch and De Deckker did on the CSIRO's RV *Southern Surveyor* was just the kind of thing that Reg would have loved: A team of scientists and students coring sediments from the sea floor; capturing samples of plankton; measuring mineral and nutrient concentrations in the water; and mapping the intricate contours of the ocean bottom—all the things that Reg had always dreamt of, and that he himself did to the best of his ability with the rudimentary equipment available to him back then.

'Reg was a lot of fun to be around. He had a great sense of humour and used to tell lots of jokes,' says Chris. 'But he was also extremely stimulating because he had all these ideas. We knew nothing about the sea floor in those days and no one had a clue what was out there. That's why everyone was attracted to him, his ideas and everything else. He was very adventurous so it was great to be involved with the things he was doing.

'He was very interested in the ecology of the sea floor. I remember working in the little boat he had before *Saori—Sea Hunt*, a little bondwood runabout with an outboard. We went out all across the Orontes bank—trying to work out where we were in the old-fashioned way of doing it—and diving down every now and again and taking a sample of the sea floor and putting it in little plastic bags.'

When Reg gained his own scuba diving credentials his sea floor sampling became more sophisticated and methodical. Between 1964 and 1969, Reg and Scoresby Shepherd sampled and recorded the sediments and sub-tidal ecology at 560 sites in Adelaide's Gulf St Vincent and Investigator Strait. This so-called 'benthic survey' was published in 1976 and was the first of its kind in South Australian waters. It is now regarded as the benchmark of sea floor flora and fauna against which cumulative changes can be measured.

'We wrote the paper in late 1975,' says Scoresby, 'and there was not much impact at the time. Twenty five years later it has suddenly became very relevant as people have become interested in the management of Gulf St Vincent. In 2000/2001, people here [at South Australian Research and Development Institute's division of Aquatic Sciences] re-surveyed the Gulf to see the way it has changed due to the impact of prawn trawling. They did a comparison between the old and the new survey and saw significant changes had taken place.'

Thanks to Reg and Scoresby's pioneering work the effect of three decades of prawn trawling and other human disturbance in the Gulf could be measured. Scientists showed that the deep-water seagrass meadows that had flourished in Investigator Strait in Reg's time and the beds of large hammer oysters that grew at Rapid Bay, have been replaced by barren sand flats: increased water turbidity due to sewage, stormwater and agricultural run-off and direct damage from prawn trawling, the likely culprits. These local extinctions and significant depletion in bryozoans and scallops in other parts of the Gulf have led to an overall decrease in habitat complexity that can affect the whole web of marine life.

It is only Reg's scuba-based survey of the ocean in the 1960s that has allowed these changes to be recorded. He and Scoresby documented intact plants and animals in their native environment at a time when other marine explorers were hauling random samples aboard or examining the jumbled contents of trawl nets to make a survey, if at all.

'It was completely novel stuff,' says Scoresby. 'He was always looking for something new. In Reg's time there was no government interest at all. Government interest has only been stimulated in the last several years with interest in the establishment of marine protected areas.'

In 2007, Reg's spirit is still palpable.

The modern home that he and Griselda built on the waterfront in 1970 for $200 000 sold for more than $2 million. The motor tunnel that he envisioned was finally built with Federal Government funding and opened in March 2000 so that thousands of motorists now traverse the 500-metre-long Heysen tunnels through the Adelaide Hills every day. Newspaper articles report the addition of a secret Ediacaran fossil site to the National Heritage List and Arkaroola's entry into the South Australian Tourism 'Hall of Fame' for accumulated awards in the category of Ecotourism. Amongst the many tourism awards Arkaroola has received in the last 17 years, this is the most appreciated by the next generation of Spriggs—Margaret and Douglas. It is a testament to their father's 40-year-old mission statement for Arkaroola: that it should be a protected place where the public could commune with nature in comfort without threatening its natural existence. Their brief had been successful. Repeat awards for 'major tourist attraction,' 'sustainable tourism,' and organised astronomical events were the icing on the cake.

In October 2007, the South Australian Museum announced commencement of the annual Sprigg lecture series in 2008. The series is sponsored by Beach Petroleum and prominent guest speakers—such as 2007 Australian of the Year, Dr Tim Flannery—will be introduced by Museum Senior Research Scientist, Dr Jim Gehling. Jim has made the study of Reg's Ediacaran fossils his life's work.

'I would have to say that the discovery and realisation of these fossils by Reg Sprigg was *by far* the most important event,' says Jim. 'It's just that—typically for Australia—it almost went unnoticed and unappreciated. Now that Reg has passed on, we truly start to appreciate the genius of this man and his courage in sticking up for his point of view.'

Belated recognition of the fundamental importance of the Ediacaran animals in the chain of animal evolution, brought Reg much satisfaction in his latter years. Being widely heralded as the discoverer of the world's most ancient animals was a mantle he accepted with humility and ease. It was a pleasing accomplishment for one who had devoted his entire life to science.

Reg survived long enough to see a proposal made to name the entire epoch of Precambrian time—542 million to 635 million years ago—the Ediacaran Period, after the abandoned silver/lead mine where he had found the first specimen. However, he was never to see it officially adopted by the International Commission on Stratigraphy in 2004. It was the first addition to the geological time scale since 1879 when the Ordovician period was inserted between the Cambrian and Silurian. Once again, geology students come exam time would be searching for new aides de memoire for remembering the order of the geological periods. 'Every Class Of Students Detests Memorizing Pointless Periods,' one science journalist suggested, as a new mnemonic to replace 'Carl's Old Shirt Doesn't Match Pete's Pants,' for remembering the order of the Cambrian, Ordovician, Silurian, Devonian, Mississippian, Pennsylvanian and Permian time periods.

There was some international debate before the title 'Ediacaran' was accepted, since fossils of Ediacaran fauna have now been found in similar rocks in Namibia, the Ukraine, Canada, the United Kingdom, Norway, Russia and China, confirming that this was a time when soft-bodied marine animals thrived around the world. Competition was even stiffer when it came time to decide where the 'Golden Spike'—the global reference point marking the beginning of the Ediacaran Period—would be located. In April 2005 it was finally hammered into 635-million-year-old rocks in South Australia's Flinders Ranges, just 100 kilometres from where the very first Ediacaran fossil was found. It was the first time a geological time period had been anchored in the southern hemisphere.

What had made Reg's Ediacara fossil finds so controversial back in 1946 was their complete unexpectedness. Even scientists who reasoned that jellyfish and other soft-bodied marine creatures *must* have filled the evolutionary niche between single-celled bacteria and animals with eyes and legs and digestive systems, never expected to find traces of them. Their bodies were surely too structureless, too amorphous, too ethereal to leave any outline in the rock.

Reg, while trying to understand his own discovery and accumulate knowledge that would help him convince others of the authenticity of his fossils, had spent much time observing jellyfish stranded on Adelaide's metropolitan beaches. He noted that the process of drying made their watery bodies quite leathery and tough and thus more amenable to preservation. The peace and stillness of Ediacara was the perfect place for their slow decay.

Once the momentum began, research into the mysterious lives of these primitive Precambrian creatures has continued unabated. Specimens of *Dickinsonia*—the circular, flattened, segmented 'jellyfish,' (really a worm) that crept over the ancient sea floor—have been found as large as dinner plates and even greater than the diameter of a car tyre. The museum has created complete dioramas of the Ediacaran sea floor complete with feathery sea pens waving in the current and quirky animals gliding across the bottom. Within their collection they have a four metre by three metre sample of fossilised seafloor preserving the largest single collection of *Dickinsonia costata* ever found. One of the cracked sections making up the complete slab could only be transported back to the city with the aid of the Channel Seven news helicopter. Five hundred and fifty-million-year-old sand ripples on the top surface shelter hundreds of tiny 'jellyfish'—looking like masses of human fingerprints—on the under surface.

Beach Petroleum continues to honour its founder and further knowledge into the mysterious lives of these and other curious Ediacaran creatures by sponsoring the museum display and providing money to fund the team's research on the Ediacara and Emu Bay fossils. The Kangaroo Island Emu Bay fossil deposit that Reg discovered has since been found to house a range of bizarre, early Cambrian marine worms and other animals of world-class importance.

Reg Sprigg would have been proud of it all. Proud that his company has the motivation to support basic science and the means to do it. Proud, too that his fossils have finally found their rightful place on the world stage. And proud that he had devoted his life to such a glorious mistress: the pursuit of knowledge—what a seductive dame.

If Reg Sprigg could have had his life all over again, he wouldn't have changed a thing. He was a man who wrung every last drop of enjoyment out of life and savoured every delicious morsel. He lived life large and inspired others to do the same.

And his genes flow on…through the blood of his daughter and his son…and his granddaughter and his grandson…two bright, enquiring young minds full of the promise and potential of a new generation.

And an unbroken link back to the man who was Sprigg.

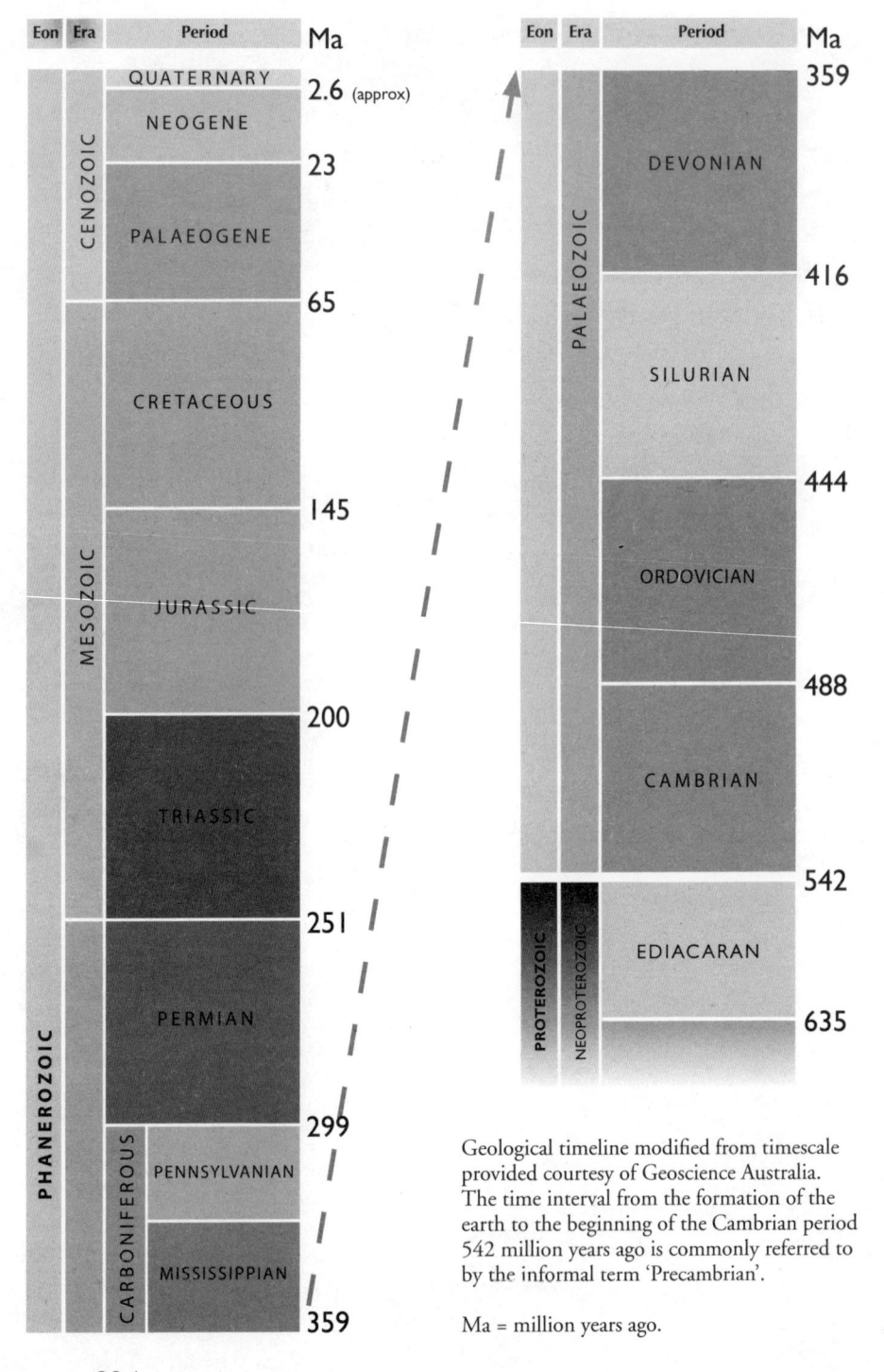

ROCK STAR

Eon	Era	Period	Ma
PHANEROZOIC	CENOZOIC	QUATERNARY	2.6 (approx)
		NEOGENE	23
		PALAEOGENE	65
	MESOZOIC	CRETACEOUS	145
		JURASSIC	200
		TRIASSIC	251
	PALAEOZOIC (PERMIAN/CARBONIFEROUS shown)	PERMIAN	299
	CARBONIFEROUS	PENNSYLVANIAN	
		MISSISSIPPIAN	359

Eon	Era	Period	Ma
PHANEROZOIC	PALAEOZOIC	DEVONIAN	359
		SILURIAN	416
		ORDOVICIAN	444
		CAMBRIAN	488
PROTEROZOIC	NEOPROTEROZOIC	EDIACARAN	542
			635

Geological timeline modified from timescale provided courtesy of Geoscience Australia. The time interval from the formation of the earth to the beginning of the Cambrian period 542 million years ago is commonly referred to by the informal term 'Precambrian'.

Ma = million years ago.

Marywadea ovata

REFERENCES

Published Material:

Ayres, Philip, *Mawson: a life*, Melbourne University Press, Carlton, Victoria, 1999.

Branagan, David, (edited by Paul Cliff) *TW Edgeworth David: A Life. Geologist, adventurer, soldier and "Knight in the old brown hat."*, National Library of Australia, Canberra, 2005.

Burchell, Dave, *One Foot in the Grave*, Heinemann, Melbourne, 1967.

Cockburn, Stewart and Ellyard, David, *Oliphant*, Axiom Books, Kent Town, South Australia, 1981.

Cooper, John, *Records and Reminiscences: Geosciences at The University of Adelaide, 1875-2000*, The University of Adelaide, Department of Geology and Geophysics, 2000.

Harrington, Malcolm and Kakoschke, Kevin, *We Were Radium Hill*, Malcolm Harrington, 1991.

Hill, PJ, De Deckker, P and Exon, NF, *Geomorphology and evolution of the gigantic Murray canyons on the Australian southern margin*, Australian Journal of Earth Sciences, vol 52, p 117-136, 2005.

Hills, Brian, *The Early Days of Hyperbaric Research in Adelaide*, in *Offgassing*, Journal of the Hyperbaric Technicians and Nurses Association, vol 33, June 2002.

Knowles, Ruth Sheldon, *The Greatest Gamblers: the epic of American oil exploration*, 2nd edition, University of Oklahoma Press, Norman, 1978.

Kutsche, Frank and Lay, Brendan, *Field Guide to the Plants of Outback South Australia*, Department of Water, Land and Biodiversity Conservation, Adelaide, 2003.

O'Neil, Bernard, *Above and Below: The South Australian Department of Mines and Energy 1944-1994*, South Australian Department of Mines and Energy, 1995.

Shepherd, SA and Sprigg, RC, Chapter 12: *Substrate, sediments and subtidal ecology of Gulf St Vincent and Investigator Strait*, in Twidale, CR (ed), *Natural History of the Adelaide Region*, Royal Society of South Australia, Adelaide, 1976.

Sprigg, Griselda, with Maclean, Rod, *Dune is a four-letter word*, Wakefield Press, Kent Town, South Australia, 2001

Sprigg, Reg, *Geology is Fun*, Reg Sprigg, 1989.

Sprigg, Reg, *A Geologist Strikes Out*, Reg Sprigg, 1993.

Sprigg, Reg C, *Early Cambrian (?) Jellyfishes from the Flinders Ranges, South Australia*, Transactions of the Royal Society of South Australia, vol 71, p 212-228, 1947.

Sprigg, Reg C, *Newly discovered submarine canyons of New Guinea and South Australia*, Nature, vol 161, p 246, 1948.

Sprigg, Reg C, *Stranded Pleistocene sea beaches of South Australia and aspects of the theories of Milankovitch and Zeuner*, International Geological Congress, report of the 18th session, part 8, p 226-237, 1948.

Sprigg, Reg C, *Early Cambrian "jellyfishes" of Ediacara, South Australia and Mount John, Kimberley District, Western Australia*, Transactions of the Royal Society of South Australia, vol 73, p 72-112, 1949.

REFERENCES

Sprigg, Reg C, *Petroleum Prospects of Western Parts of Great Australian Artesian Basin*, Bulletin of the American Association of Petroleum Geologists, vol 42, p 2465-2491, 1958.

Sykes, Trevor, *The Money Miners: Australia's Mining Boom 1969-70*, Wildcat Press Sydney, 1978.

Tanner, JE, *Three decades of habitat change in Gulf St Vincent, South Australia*, in Transactions of the Royal Society of South Australia, vol 129, p65-73, 2005.

Wilkinson, Rick, *A Thirst for Burning: The Story of Australia's Oil Industry*, David Ell Press, Sydney, 1983.

Unpublished Material:

University of Adelaide Archives:
Series 200, Registrar's Correspondence Files [Dockets]; Items 1938/251, 1944/42 and 1955/115.
Series 618, Dr RC Sprigg Memorial Ceremony.

Correspondence between Sir Mark Oliphant and Dr RC Sprigg, Oliphant papers, Barr Smith Library, University of Adelaide, MS 92 04775p/Series 3.

National Archives of Australia: Australian Security Intelligence Organisation; Series A6119, personal file; Item 1866, 1949-1960.

State Library of South Australia:
OH 89, Department of Mines and Energy Oral History Program 1980-1993, transcript numbers 2, 5, 8, 18, 21, 22, 25, 32 and 34.
OH 485, Strzelecki Track Oral History Program, 1995, transcript number 6.

Mawson Collection, South Australian Museum: Correspondence file 63DM.

Reginald Claude Sprigg private family archives

Richard Grenfell Thomas private family archives

State Records of South Australia GRS/6038 Department of Mines, General correspondence dockets; sub-series 00017 (1915-1987) and sub-series 00014 (1915-1972).

Phyllozoon hanseni

INDEX

* photos underlined

Adelaide Geosyncline: 96-97, 106-107
Adelaide River: 114, 201
Adelaide Technical High School: 10-11
Admella: 5-6, 299
airborne scintillometer: 89-90
Alderman, Arthur: 12, 18, 97, 110
Alf Flat, *see* Salt Creek
Ampol, *see* WAPET
anticline: 132, 135, 136, 137, 170, 173, 196
ANZAAS (Australian & New Zealand Association for the Advancement of Science): 49, 74, 100, 284
APEA/APPEA (Australian Petroleum Production & Exploration Association): xvi, 260-261, 310-311, 315

Archaeocyatha: 11, 48, 283
Arkaroo legend: 215
Arkaroola Station: 19, 32, 34, 214, 216-217
Arkaroola Wildlife Sanctuary: xvii, 217-221, 226, 234-236, 239-240, 246-250, 252, 254-255, 257, 260, 266, 270, 275-276, 279-282, 281, 285-287, 300, 306, 309-310
  accommodation: 220-221, 234
  observatory: 280, 282
  Ridgetop Track: 236, 237, 238
  tourism awards: 320
ASIO (Australian Security Intelligence Organisation): 73-75, 91
Asteroid Sprigg: xvi
Astrofixing: 102, 163-164
Atomic Energy Commission, UK/US: 54, 56, 64, 71, 97

328

INDEX

Australian Amateur Mineralogist: 151
Australian Association of Scientific Workers: 74
Australian (Petroleum) Scouting Service: 134, 151

Bass Strait oil field: 173, 194, 203, 209, 267, 313
Beach Petroleum: 221, 247, 250, 254, 313-314, 320, 322
  incorporation: 152
  Simpson Desert operations: 153, 168-169, 254, 256, 267-269, 272
  South East ops: 153, 168, 170, 173, 193-194, 254, 267-269, 272, 313
  St Vincent Basin ops: 170, 187, 203-204
  Turkey ops: 222-224, 227-233, 246, 254
  maiden dividend: 268
  takeover by Claremont: 269-274, 278, 290
  takeover by IRL: 271-275
  court case vs IRL: 274, 290-291, 303-304, 311-312
  Delhi takeover: 312-313
Bodalla oil field: 268-269, 290, 311
Bonython, John: 109, 112, 115-116, 120, 127, 129-131, 137-138, 140, 146, 148, 196, 287, 290
Bristowe, Robert: 109, 112, 115-116, 120
Brunnschweiler, Rudi: 113, 115-116, 128-129, 135-136, 142
Burchell, Dave: 179-180, 185-186, 205-208, 318

Claremont Petroleum: 269-274, 278, 290
coastal bitumen: 170-174, 193-194
Colson well: 254-256, 267, 272
Combined Development Agency, UK/US CDA: 124, 201
coorongite: 170-171
Crockers Well: 90-91

Curtin, John, Prime Minister: 26, 36

David, Edgeworth: xvi, 27, 43-45, 51-53, 59
davidite: 27, 43
Day, Patricia: 36, 39-40, 292-293, 301-303
De Deckker, Patrick: 316, 318
Delhi Taylor Oil Corporation: 137-143, 146-148, 150, 196-197, 312
Department of Mines, SA: 12, 26, 50, 62, 67, 70, 75, 82, 92, 98-101, 104, 106, 112, 115, 129, 132, 139, 142, 195-197, 283, 292, 297
  Geological Survey: 25, 37-38, 64, 113
Dickinson, Ben: 25-26, 28, 32, 35, 37, 50-51, 53-61, 63-64, 66-67, 70-72, 74, 82, 91-92, 95, 97, 99, 103-104, 106-107, 110-111, 126
Divana-1 well: 223-224, 228-231
diving chamber: xvii, 178, 183-186, 184
Dodwell, George: 281
Dolling, Ngaire: 79-81

Ediacara mine: 30, 39-40, 45, 47-49
Ediacaran fossils: 50, 52-53, 62, 174, 283-284, 320-322
  Beltanella: 1, 51
  Dickinsonia costata: 51, 85, 322
  Ediacaria flindersi: 47, 50
  Mawsonites spriggi: 169, 284
  Parvancorina minchami: 193, 284
  Spriggina floundersi: xvi, 213, 282, 284
Een, Patricia, *see* Day, Patricia
eurypterid: 41-42, 45
Exoil: 236-238

Francis, Dorothy: 199-201
French Line: 198-199
Frome Broken Hill Group (Frobilco): 127-128, 137, 141, 146, 148, 195, 267
Fuller, Michael: 271, 273, 290, 303, 311

329

Gaia: 278, 307
Gehling, Jim: 320
Geltwood Beach: 170, 172-173, 193, 299
Geosurveys: 116, 119-120, 128-129, 132-134, 142-<u>143</u>, 146-147, 150-153, 158-159, 164, 169, 173, 177-178, 180, 183, 185-186, 195, 199-200, 203, 221, 223-224, 226, 229, 241, 244-245, 249, 287, 294, 312
  formation: 110-112
  takeover by Beach: 246
  reunion: 299-302
Germein, Captain John (RCS great, great-grandfather): 3-4
  John II (great-grandfather): 4, 299
  Samuel (great, great-uncle): 4, 299
  Benjamin (great, great-uncle): 4-6, 299
  John III (grandfather): 2-3, 5
  Pearl (mother): 1-3, 7-8, 114, 252-253
Gidgealpa-2 well: 195-197
Giles Complex: 251
Glaessner, Martin: 96-97, 116, 282, 284
gravity survey:
  Simpson Desert: 153-154, 160, 163, 169
  Gulf St Vincent: 183, 187, 190, 203, 243
  Cape York Peninsula: 241-243, <u>242</u>
  Turkey: 223-224
Great Artesian Basin: 107, 116-118, 120, 127, 129-130, 135, 137, 139, 141, 146, 149, 196
Greenwood, WB (William Bentley): 29-31, 214
  Gordon (Smiler): 19, 30, 32, 119, 214, 216, 221
  Bentley: 32, <u>34</u>, 214, 216, 235

Halley's Comet: 280
Howchin, Walter: 10-11, 51, 59, 277

Independent Resources Ltd Group: 271-275, 290, 304, 311-312, 314

Innamincka-1 well: 142, 148-<u>149</u>, 174, 195, 311
International Nickel Company: 119, 151, 201, 209, 250-251

Jack, R Lockhart: 97, 131-132, 135-136, 138
Johns, Keith: 13-14, 17, 52, 58
Johnson, Jim (Hoppy): 101-103

Kangaroo Island: xvi, 36, 48, 61, 95-96, 175-176, 187
  Emu Bay: 95, 322

Laws, Bob: 103, 134, 227, 232-233, 241, 254
Le Messurier, Hugh: 188
Levorsen, AI (Lev): 130-131, 137

Madigan, Cecil T: xvi, 9, 11-15, <u>12</u>, 17, 21, 38, 52, 59, 94-95, 154, 168, 277
mapping, geological: 25, 32, 38-39, 67, 71-73, 93, 142
Marchant, Olive: 85-86
Mawby, Maurie: 76-77, 109, 127-128, 141
Mawson, Douglas: xvi, 9, 11, 15-24, <u>20</u>, 27, 29-31, 59, 97, 104-106, 110, 173-174, 214-215, 217, 220, 235, 265, 277, 281
  relationship with Madigan: 12-14, <u>12</u>, 17, 21, 38, 52
  fossils: 40-41, 43-45, 49-51
  SANTOS: 112, 116
  namesakes: 119, <u>169</u>, 234, 238, 247, 284, 308
  Paquita: 20, 234
McBriar, Maude: 74
McEwin, Lyell: 82, 128, 139
Menzies, Robert: 67-70, <u>68</u>
Middleton, Rennie: 86, 98, 152
  Roslyn: 86-87
Milankovitch cycles: 298-299, 315-316
Mincham, Hans: 283-284

INDEX

Mount Davies: <u>102</u>, 105, 116, 119, 151, 195
Mount Gee: 215-216, 297-298
Mount Lofty road tunnel: 296-297, 320
Mount Oliphant: 280
Mount Painter: 54, 56-57, 68-69, 92, 220, 235, 238, 250, 264, 309-310
  radium discovery: 29-31
  wartime uranium project: 26, <u>31</u>-33, 35-36, 40, 53, 72, 214-215, 236
Mundrabilla meteorite: 294

Naturelle Jewellery: 151-152, 226, 241
Nelson, Reg: 291, 303, 312, 314
NERDDC (National Energy Research Development & Demonstration Council): 256, 260
nickel: 105, 113, 116, 119,151, 201, 209-211, 250
Nickel Mines of Australia: 116, 151, 210-212, 247, 250
North Paaratte-1 well: 267, 269, 272
North-west survey: 100-106, <u>102</u>

Oliphant, Mark: 54-56, 64, 71, 82, 264-266, 269, 271, 277-280, 287-288, 294-295, 300-301, 303, 307, 310, 315
Otway Basin: 113, 193-194, 254, 267, 272, 313

Paralana Hot Springs: 235-236
Parkin, Lee: 15-16, 92, 98-99
Paterson, Grace: 76, 78, 81, 83
  Griselda, *see* Sprigg, Griselda
Petroleum Search Subsidy: 246
Playford, Tom, Premier: 35, 37, 50-51, 53-54, 57, 59, 61, 64, 67-71, <u>68</u>, 75, 82-83, 86, 90, 97, 99, 122, 124-125, 129, 139, 142-143, 145-<u>149</u>, 174-175, 197, 235, 296
Poeppel, Augustus: 94, 163
  Post: 163-<u>165</u>

Corner:163, 166
Poseidon nickel boom: 209-211, 247
Priess, Wolfgang: 106

radium: 27, 29-31
Radium Hill: 26, 28-29, 31, 36-37, 55-59, <u>57</u>, 63-64, 67-69, <u>68</u>, 89-90, 93, 150, 283
  discovery: 27
  mine: 70-71, 73, 79, 82-83, 92, 97, 122, 124-125, 202-203
  security: 73, 75, 97-99
  township: 70, 73, 85-<u>88</u>, 122-124, 202-203, 300
Radium Ridge: 30
Radus, Jonas: 181, 187, 190-191, 243
Rough Range: 108, 113, 115, 118, 120, 132, 141, 148, 195
Royal Society of South Australia: 43, 45, 51, 105, 173, 266, 283-284

Salt Creek/Alf Flat: 170-171
SANTOS (South Australia Northern Territory Oil Search): xvii, 120, 127, 142, 274, 287, 311-312
  formation: 112-116
  prospectus: 114-115, 117-118
  Great Artesian Basin acreage: 116-117, 128-132, 135, 137, 142-143, 146, 148, 150, 195-197, 311
  first oil well, *see* Wilkatana
  negotiations with Delhi: 137-141
  gas discovery, *see* Gidgealpa-2 well
  Cooper Basin operations: 208, 250, 288-290
  maiden dividend: 269
  oil discovery: 289-290
Saori: 178, 183, 185, 204, 206-207, 246, 319
  launch: 174-<u>176</u>
  voyages: 187, 189-193, 205, 299
  marine seismic: 203-204
  demise: 250-252

331

scuba diving: 177-182, <u>184</u>-192, 205-208, 318-319
Shaw, Jean: 272-273, 290-291, 314
Shepherd, Scoresby: 189-192, 204-205, 319-320
Simpson Desert crossing, Colson: 154
  Madigan: 94, 154, 168
  Sprigg: 153-168, <u>157</u>, <u>167</u>, 197-198
South Australian Museum: 9-10, 189, 284, 294, 320, 322
Specimen Minerals: 134, 151, 226
Sprigg, Samuel Augustus (RCS grandfather): 2-3
  Claude Augustus (Pop) (father): 2-3, 6-7, 9, 114, 134, 151, 172, 252-253
  Pearl, (mother) *see* Germein, Pearl
  Reginald Claude, childhood: 1-3, 6-9, 134
    student: 10-17, 19, 21-24, 38, 42
    relationship with Mawson: 11, 15, 19, 21-24, 33, 38, 40-41, 43, 45, 104-105, 119, 173-174, 215, 217, 265, 277, 284
    university cadet: 15-17
    secondment to Dept Mines: 25-26, 29
    first marriage: 36, 39-40, 58, 62, 75, 282, 292
    government geologist: 37, 106, 110, 113, 172
    overseas tour: 1948: 61-67, 108, 283
    Griselda courtship: 64-<u>65</u>, 75-<u>80</u>
    ASIO target: 73-75, 91-92
    fossil discoveries: xvii, 40-43, 45, 47-51, 53, 96, 174-175, 282-284, 287, 292, 320-323
    marine exploration: 60-61, 173-178, 187, 204-208, 298-299, 315-320, <u>317</u>
    uranium exploration: 28, 32-35, 55-58, <u>57</u>, 67-<u>68</u>, 70-71, 89-91, 100, 125-126, 201, 203

oil exploration: xvi, 67, 93-94, 107, 116-118, 126-132, 135-137, 148-150, 152, 172-173, 193-194, 196-197, 223, 227-231
debt: 240, 246-250
friendship with Oliphant: 264-266, 269, 271, 277-280, 287-288, 294-295, 300, 303, 307, 310
memoirs: 276-277, 286-287
scientific publications: 50-53, 59, 61, 137, 146, 173, 283, 298, 316, 319
illness: 232, 253, 256-260, 262, 267, 270, 288, 300-302, 305-307
awards & tributes: xv, 262-<u>264</u>, 266, 277, 309-311
death: 307-309
D'arcy (brother): 1, 252
Connie (sister): 1, 252
Griselda (wife): 95, 106, 108, <u>126</u>, 129, 143-144, 174-175, 182, 211-212, 252-254, 266, 270, 293, 296, 299-300, 302-306
  life in Scotland: 64-<u>65</u>, 75-77, 259, 305, 309
  wedding: 78-81, <u>80</u>
  family life in Adelaide: 82-83, 92-93, 100, 111, 118-121, 217-<u>219</u>, <u>218</u>, 224-226, 256-260, 320
  Radium Hill: 84, 87-90, <u>88</u>, 122
  Simpson Desert: 153, 155-159, 161-163, 166-168, 198-199
  Arkaroola: 214-217, 220-221, 226, 233-238, 240, 247-249, 255, 275-276, 286, 301
  correspondence in Turkey: 222, 224, 227-230
  after Reg: 308-311, 320
children: 119-121, 143-<u>144</u>, 153,

INDEX

155-158, <u>157</u>, 161, <u>165</u>, 198, 215, 217, 225, 236-238, 248, 300, 305, 310, 320
  Margaret (daughter): 100, 118, 226, 303
  Douglas (son): <u>126</u>, 162, 220, 256, 279-280, 309, 314
Sprigg Canyon: xvi, 316-318, <u>317</u>
spriggite: xvi
Strzelecki Track: 142, 146-147
submarine canyons: xvi-xvii, 60-61, 175, 177, 316-318, <u>317</u>

Teesdale-Smith, Ngaire, *see* Dolling, Ngaire
Teichert, Curt: 49-51, 62, 282
Thomas, Richard Grenfell: <u>12</u>, 18-19, 135, 220
Thomson, Clyde: 178, 240-248
torbernite: 30, 55, 114
trilobite: 41-42, 95-96
Turkey: 222-225, 227-234, 240, 246, 253-254, 277

University of Adelaide: xvi, 12-13, 15, 24, 27, 29, 38, 54, 95-96, 110, 188, 265, 282, 309
  Geology Club: 25, 42-43, 45, 100
  RCS doctoral thesis: 97, 106-107, 263, 287
  nickel shares: 211
uranium: xvii, 25, 29, 36-38, 54, 61-64, 67, 70-71, 75, 89-93, 98-100, 105, 108-109, 113-114, 116, 124-125, 201, 236, 265
Uranium Development & Prospecting: 114, 201

Vial, Allan: 101-103, 116, 132, 241
von der Borch, Chris: 316, 318
von Sanden, Darby: 153, 159-163, 166-<u>167</u>, 178, 180, 182-<u>184</u>, 186, 191-192, 203, 205-208, 223, 241, 244-245, 318

WAPET (West Australian Petroleum)/ Ampol: 108, 114-115, 120, 141, 208
Ward, L Keith: <u>25</u>-26, 97, 107-108, 115-116
Warman, Peter: 179-182, 185
Weeks, Lewis: 194
  medal: 266
Whittle, Alick: 58, 109
Wilkatana: 94, 115-116, 118, 120, <u>126</u>-130, 135, 148, 311
Wilson, Bruce: 101-103, 109, 130, 133, 293-294
  Betty: 133, 153
Wopfner, Heli: 128-129, 132-133, 135, 142, 144-145, 148, 195-197, 233
Wright, Trish: 163-<u>164</u>

Zinc Corporation: 28, 76-77, 94, 109, 127

About the Author

Kristin Weidenbach is the author of *Mailman of the Birdsville Track: the Story of Tom Kruse*.

Kristin is a PhD immunologist who switched to a writing career after completing postdoctoral research at Stanford University in California. She is the author of several academic publications and the recipient of an award from the International Association of Forensic Sciences.

As a science writer Kristin has written for *Science* magazine and for publications from Harvard and Stanford Medical Schools. Her magazine articles have been awarded prizes from the *Council for the Advancement and Support of Education* and from the *Association of American Medical Colleges*. She has also written for *Outback* magazine and *Australian Geographic*.

After seven years living in America, Kristin currently lives and writes in Adelaide. *Rock Star: the story of Reg Sprigg*, is her second book.